Casting Protocols for the Upper and Lower Extremities

THE REHABILITATION INSTITUTE OF CHICAGO
PUBLICATION SERIES
Don A. Olson, PhD, Series Coordinator

Spinal Cord Injury: A Guide to Functional Outcomes in Physical Therapy Management

Lower Extremity Amputation: A Guide to Functional Outcomes in Physical Therapy Management, Second Edition

Stroke/Head Injury: A Guide to Functional Outcomes in Physical Therapy Management

Clinical Management of Right Hemisphere Dysfunction, Second Edition

Spinal Cord Injury: A Guide to Functional Outcomes in Occupational Therapy

Spinal Cord Injury: A Guide to Rehabilitation Nursing

Head Injury: A Guide to Functional Outcomes in Occupational Therapy Management

Speech/Language Treatment of the Aphasias: Treatment Materials for Auditory Comprehension and Reading Comprehension

Speech/Language Treatment of the Aphasias: Treatment Materials for Oral Expression and Written Expression

Rehabilitation Nursing Procedures Manual

Psychological Management of Traumatic Brain Injuries in Children and Adolescents

Medical Management of Long-Term Disability

Psychological Aspects of Geriatric Rehabilitation

Clinical Management of Dysphagia in Adults and Children, Second Edition

Cognition and Perception in the Stroke Patient: A Guide to Functional Outcomes in Occupational Therapy

Spinal Cord Injury: Medical Management and Rehabilitation

Clinical Management of Communication Problems in Adults with Traumatic Brain Injury

Rehabilitation of Persons with Rheumatoid Arthritis

Functional Rehabilitation of Sports and Musculoskeletal Injuries

Casting Protocols for the Upper and Lower Extremities

Paula Goga-Eppenstein, MS, PT
Level II Physical Therapist
Loyola University Medical Center
Maywood, Illinois

Judy P. Hill, OTR/L, BS, OT
Administrative Clinical Director
Spinal Cord Injury and Amputee
  Clinical Operating Group
Rehabilitation Institute of Chicago
Chicago, Illinois

Puliyodil A. Philip, MD
Associate Director
Brain Injury Program
Attending Physician
Rehabilitation Institute of Chicago
Chicago, Illinois

Mersamma Philip, MD
Attending Physician
Physical Medicine and
  Rehabilitation
VA Chicago Health Care System,
  Lakeside Division
Chicago, Illinois

Terry Murphy Seifert, BS, PT
Marionjoy Rehabilitation and
  Paradigm Restorative Services
Wheaton and Naperville, Illinois

**Audrey M. Yasukawa, MOT,
  OTR**
Pediatric Resource Clinician
Rehabilitation Institute of Chicago
Chicago, Illinois

With

Patricia A. Lee
Consulting Editor and Writer
Past President, American Medical Writers Association
Greater Chicago Area Chapter
Chicago, Illinois

AN ASPEN PUBLICATION®
Aspen Publishers, Inc.
Gaithersburg, Maryland
1999

Day Shore

The author has made every effort to ensure the accuracy of the information herein. However, appropriate information sources should be consulted, especially for new or unfamiliar procedures. It is the responsibility of every practitioner to evaluate the appropriateness of a particular opinion in the context of actual clinical situations and with due considerations to new developments. The author, editors, and the publisher cannot be held responsible for any typographical or other errors found in this book.

All photos in this publication were taken by Movco Media, copyright © Rehabilitation Institute of Chicago.

Library of Congress Cataloging-in-Publication Data

Casting protocols for the upper and lower
extremities / Paula Goga-Eppenstein…[et al.].
p.   cm. — (The Rehabilitation Institute of Chicago
publication series)
Includes bibliographical references and index.
ISBN 0-8342-0763-X
1. Plaster casts. Surgical. 2. Arm. 3. Leg.
I. Goga-Eppenstein, Paula. II. Series.
[DNLM: 1. Casts, Surgical—Muscular Diseases.
2. Arm—Muscular Diseases.
3. Leg—Muscular Diseases. WO 170 C352 1999]
RD114.C37   1999
617.5′8—dc21
DNLM/DLC
for Library of Congress
99-31756
CIP

Orders: (800) 638-8437
Customer Service: (800) 234-1660

About Aspen Publishers • For more than 35 years, Aspen has been a leading professional publisher in a variety of disciplines. Aspen's vast information resources are available in both print and electronic formats. We are committed to providing the highest quality information available in the most appropriate format for our customers. Visit Aspen's Internet site for more information resources, directories, articles, and a searchable version of Aspen's full catalog, including the most recent publications:
http://www.aspenpublishers.com

Aspen Publishers, Inc. • The hallmark of quality in publishing
Member of the worldwide Wolters Kluwer group.

Editorial Services: Kathy Litzenberg
Library of Congress Catalog Card Number: 99-31756
ISBN: 0-8342-0763-X
Printed in the United States of America
1 2 3 4 5

10/12/2000

Table of Contents

Foreword

Contracture, spasticity, and other abnormalities of muscle tone and position are frequent concomitants of many disabling conditions. They can limit or reduce function, delay or impede rehabilitation efforts, and cause or exacerbate both disability and medical morbidity. These problems can contribute to the occurrence of functional loss or medical complications by themselves, or they can lead indirectly to pain, discomfort, and additional disability. The goal of the rehabilitation professional to maximize function in patients with disabling conditions requires the knowledge and skills necessary to prevent, treat, and minimize the impact of these important problems. For many patients, contracture is their most important limiting factor. For some, it is a relatively minor annoyance. Even when strength and tone are relatively functional, contracture can interfere with the performance of specific activities. Pain and disfigurement are other serious consequences of these problems.

Successful treatment of contracture requires a thorough knowledge of prevailing therapeutic interventions, superior clinical judgment to assess and implement the appropriate and timely use of available modalities, a high level of competency necessary to perform the requisite clinical skills properly, and creativity, perseverance, and dedication to improve function and reduce disability. A variety of techniques are available to treat contracture and to minimize its impact.

One of these techniques, casting, is reviewed extensively in this book. In particular, the book features attention to the practical details involved in performing the necessary techniques properly. The characteristic that is unique about this text, however, is that the focus is not only on the development of clinical competency in performing the technique itself (the protocols), but also on the *theoretical framework* that forms the basis for the procedure. This is important because gaining an appreciation for the conceptual basis of the clinical skill helps to enhance the clinical proficiency of the professional. Studying the theoretical framework not only makes learning about the technique more intellectually stimulating, but it also facilitates more appropriate application of the procedure in a variety of circumstances that might otherwise have been missed by the clinician. Therefore, although the practical considerations are essential to the proper performance of the casting procedure, using this book as only a "how-to" reference guide would be selling the text—and the clinician's opportunity for clinical skill development—too short. The reader is encouraged to understand the theoretical and conceptual basis of casting as a means of facilitating the application of the method for a variety of situations.

Judy Hill, Audrey Yasukawa, and the team that they have assembled from the Rehabilitation Institute of Chicago are highly skilled clinicians whose extensive experience includes not only performing the techniques described in this book, but also *teaching* the methods and their applications to others. The authors have drawn upon their many collective years of performing the casting techniques in a variety of settings, their experience in mentoring clinicians by demonstrating the technique, and their recognition of the importance of gaining a thorough understanding of the conceptual background that underlies the skill in order to provide a clear, practical, and complete explanation of the applications and methods of casting techniques. Although the clinician is expected to gain considerable benefits from the authors' expertise and explanations, it is likely that the real winners will be the patients whose outcomes will be enhanced by their applications.

Elliot J. Roth, MD
The Dr. Paul B. Magnuson Professor and Chairman
Department of Physical Medicine and Rehabilitation
Northwestern University Medical School
The Donnelley Senior Vice President and Medical
Director
Rehabilitation Institute of Chicago
Chicago, Illinois

Preface

This book guides rehabilitation professionals, primarily occupational and physical therapists, in the use of casts to manage upper and lower extremity contractures and hypertonicity. It reflects the authors' many years of experience with the therapeutic applications of casts and their knowledge resulting from interactions with other therapists and physicians. The initial casting techniques were introduced at the Rehabilitation Institute of Chicago by Christine Chapparo, OTR/L, who brought them from her experience at Rancho Los Amigos Hospital in Downey, California. Interactions with Beverly Cusick, MS, PT, influenced the use of lower extremity casts, particularly casts for children. Additional techniques and types of casts were developed to refine the basic casts over the years.

Experience has guided us in developing new types of casts as well as in refining candidate selection and more accurately predicting goals. We've also learned to use other techniques in conjunction with casting.

Even though casting can often reduce an impairment, the application to function and the reduction of disability and handicap does not always follow. It is our hope that this book conveys our experience in a practical and understandable manner, guiding the reader in the rationale for applying casts, goal setting, assessment, and actual cast application.

We also caution the reader that cast application as described does carry some risk. Skin and vascular problems have been attributed to improper cast application. We do not recommend that practitioners without any supervised hands-on experience in applying casts do so solely with the instructions provided in the following chapters. Instead, we recommend taking courses about casting and collaborating with other professionals or technicians who have experience with cast application. Cast technicians, some orthotists, orthopaedic specialists, and other therapists with experience can provide the mentoring to accompany this book.

Acknowledgments

We acknowledge: Dr. Henry Betts for creating an environment of inquiry and clinical excellence at the Rehabilitation Institute of Chicago; Dr. Don Olson for his continuing encouragement and support; Fred Schneider for his help during the early stages of this book; also Beverly Cusick for her significant contributions to the use of casting in rehabilitation and Christine Chapparo for bringing casting techniques to the Rehabilitation Institute of Chicago from Rancho Los Amigos Hospital in Downey, California.

Finally, we thank our families for allowing us to take the opportunity to fit one more important project into our busy lives:

Audrey—husband Danny and son Miles Faith
Paula—husband David Eppenstein
Terry—husband John Seifert and children Molly and Jack
Judy—Rodney Estvan and daughters Ali and Emma Estvan Hill
Pat—husband Ray Jones

CHAPTER 1

Theoretical Background and Rationale for Cast Intervention

Paula Goga-Eppenstein, Judy P. Hill, Terry Murphy Seifert, and Audrey M. Yasukawa

Contractures and hypertonicity with associated limitations of passive and active range of motion are impairments frequently addressed in the rehabilitation of individuals with neuromuscular, neurological, and some soft tissue disorders. These impairments in range of motion and movement are thought to interfere with functional use of the extremities and task performance. Contractures can occur in muscles and other soft tissues due to prolonged immobility in a shortened position.[1,2] The prolonged immobility and resulting adaptive changes in the muscle can occur from an imbalance of muscle power or prolonged, fixed postures due to central nervous system damage. They are thought to occur more quickly in innervated than denervated muscle.

Motor disorders in which contractures and hypertonicity are often observed include quadriplegia and paresis, hemiplegia and paresis, and paraplegia and paresis. Frequently, these disorders are associated with spinal cord injury, brain injury, stroke, and cerebral palsy. In other conditions, such as burns, contractures result from initial damage to soft tissues and muscles, as well as subsequent scar formation.

Traditional rehabilitation programs have included techniques such as passive and active range of motion exercises, splinting or orthotic intervention, and static and prolonged stretch to prevent and correct contractures. Neurophysiological treatment techniques such as proprioceptive neuromuscular facilitation and neurodevelopmental treatment have been used to inhibit the effects of hypertonicity to mobilize and elongate shortened muscles. Some of these techniques focus on biomechanical elongation of the shortened tissues and others on influencing the neurological system through the muscle spindle afferent feedback mechanism. For a variety of reasons, these techniques have not always met with success in reversing contracture and restoring more balanced use of the extremities. Most of them, if effective during the treatment in relaxing and elongating the muscles, do not maintain their effectiveness outside of treatment. With immobilization, joints can become stiffer and resist attempts at passive elongation. In animal limbs immobilized for longer than 16 days, the force needed to stretch the limb was four to six times greater than the force required to stretch mobile joints.[3,4] Passive range of motion, if performed too aggressively, may overstretch tissues, causing them to tear and scar. Scarring further limits plasticity and extensibility.[5] Kottke et al.[6] and Bell[7] refer to the effectiveness of prolonged stretch in preventing and correcting contractures in connective tissues. Bell specifically refers to the use of casts to achieve the prolonged stretch. Kottke et al.[6] discuss the properties of connective tissue that contribute to contracture formation. These include a tensile force with a tendency toward shortening unless stopped by an opposing force. They suggest that low-load, long-duration stretch is the optimal method for stretching shortened connective tissue. This method decreases the risk of tearing soft tissue and increases residual elongation, optimizing the plastic component of connective tissue. A low load of force and a long duration of stretch will also more permanently realign collagen fibers into a more parallel orientation. This parallel orientation minimizes tissue tearing. Low-load, long-duration stretch results in a more permanent elongation of connective tissue.[8,9]

In many of the conditions requiring contracture and spasticity management, there is impairment or absence of opposing force due to muscle weakness or total loss of voluntary muscle contraction. Splints have been thought to provide elongation and inhibition by positioning the limb in a static position with the muscles and soft tissues on stretch. Splints, however, do not provide as prolonged a stretch as a cast because they are usually removed several times a day. During periods of removal, the tissues are allowed to resume their shortened position. Splints also have a biomechanical disadvantage compared with casts in that they usually are applied to a single surface, dorsal or volar, anterior or posterior, allowing more movement with the splint in place than with a full-circumference cast.

One mechanism underlying the adjustment in length of muscle has been offered by Williams and Goldspink.[1] They discuss a decrease in the number of sarcomeres, the longitudinal segments in striated muscle, in muscles maintained in a shortened position and an increase in the number of sarcomeres in muscle maintained in a lengthened position. These changes result in an adjustment in muscle extensibility, a loss of flexibility in muscles maintained in a shortened position. A sharper and steeper passive length–tension curve than normal results, demonstrating the loss of flexibility as the muscle more quickly develops maximal tension with less elongation.[10–12]

When muscle spasm is present, a sharp increase in passive tension is generated at shorter muscle lengths in immobilized, compared with mobile, limbs.[11] When peak amplitude of tension occurs closer to the point of immobilization, the muscle is at risk for fiber tearing and scarring. Repositioning of a joint following immobilization facilitates recovery of sarcomere number and muscle length–tension relationships. When repositioned and maintained at intermediate positions, passive length–tension curves are between those for muscles in a shortened position and those of fully mobile muscles.[12] In muscle maintained in a lengthened position, with sarcomeres added on, the passive length–ten-

sion curve approaches that of a fully mobile muscle, developing throughout the range. Sarcomere number increases or decreases in relationship to the angle the joint is positioned in and the position of the attached muscle. Thus maximal contractile tension and maximal rate of shortening of the muscle are obtained.

Sarcomere decrease in hypertonic muscles is also exaggerated. Sarcomere number decreases 25% to 40% within 14–28 days in immobilized limbs and increases to a 45% loss in that amount of time in hypertonic muscle.[10,11] The changes in sarcomere number also occur more rapidly when abnormal tone is present, with changes noted in 2 to 4 days compared with 5 days in normal muscle.[13] With immobilization, there is increased connective tissue proliferation. Adhesions can occur by day 15. There is increased cross-link formation, which results in decreasing water and glycosaminoglycan content, increasing soft tissue stiffness.[4]

The above properties of muscle, connective tissue, and soft tissues suggest a number of implications relevant to cast intervention for managing contracture and hypertonicity. A hypertonic limb might rapidly exhibit alterations in muscle length and loss of extensibility. Repositioning with a cast might allow for addition of sarcomeres and alteration of the length–tension curve, especially if the cast does not put the muscle and other tissues on maximal stretch, which could cause further tissue damage. The importance of timing is also suggested, as the cast itself is an immobilizing device. Leaving a cast in place for more than 4 to 5 days, particularly a full-circumference cast, might contribute to an increase in the detrimental effects of immobilization.

By the late 1970s and 1980s therapists and physicians began reporting the results of immobilization with a series of plaster casts applied to spastic extremities with soft tissue and muscle contractures.[7,14–21] As early as 1949, Brennan[22] suggested that immobilization in splints could alter abnormal joint position and in some cases result in an improvement in function in poststroke, hemiplegic patients' upper extremities. Casts were thought to be more effective than other treatment techniques in altering muscle length and soft and connective tissue extensibility because they provide the low-load, long-duration stretch indicated, allowing for tissue expansion and sarcomere addition. Also, casts were postulated to be more effective than other techniques in reducing hypertonicity through interruption of muscle spindle excitation.[19,23]

The influence of casts on hypertonicity continues to be speculative, with research that documents the effects in human subjects lacking. It is known that muscle contractibility is regulated in part via the muscle spindles that lie parallel to muscle fibers in striated muscle. The muscle spindles help to regulate the excitability of the muscle and are responsive to changes in muscle length and tension. They respond variably to the velocity of the changes in length. Spasticity has been described as associated with hyperexcitability of the stretch reflexes and muscle spindle activity. It is possible that these muscle spindle influences, resulting in excitation of the muscles, are interrupted with a cast in place. The cast could prevent changes in length of the muscle, eliminating the excitatory input of the muscle spindle.

Primary (group Ia) endings are highly sensitive to the rate of change of muscle length. They are sensitive to the dynamic phase of the stretch. Research has demonstrated that muscle spindle Ia afferent activity declines with sustained muscle stretch. Passive ankle dorsiflexion reduces Ia afferent discharge and the amplitude of the H-reflex. Pressure over the Achilles tendon also reduces the amplitude of the H-reflex, although not to the same degree as dorsiflexion.[24,25] If Ia synaptic activity is reduced following stretching and pressure over the Achilles tendon, then no further force output should be present to resist newly imposed changes in muscle length, and the degree of muscle tone should diminish. With casting, prolonged passive dorsiflexion could be initiated and result in a decrease in tone through this process.

Group II muscle afferents, which excite flexor motoneurons and inhibit extensor motoneurons, are excited by a steady or newly maintained stretch length.[26,27] Studies on the effect of stretch support a progressive decline in the H-reflex with a resultant inhibition of the extensor (dorsiflexor) motoneurons.[26,28] Decreased motoneuron input due to decreased firing of the Ia afferents and activation on non–spindle group II afferents

may decrease muscle tone so range of motion can improve.

There is also suggestion that neutral warmth might help to inhibit excitability and promote relaxation of muscles. King[23] suggested that this neutral warmth with the cast in place may contribute to relaxation of spastic muscles.

Whether casts actually alter hypertonicity is still debated. Gossman et al.[2] reported that the alteration in muscle length and atrophy of muscle due to immobilization may be responsible for the observed changes in muscle tension and elasticity following casting. The relationship between joint stiffness and reflex gain is also discussed by Lee et al.[29] They found lack of support for increases in stiffness in spastic muscles compared with the contralateral limb and suggested that stiffness, if found, is really related to mechanical properties of muscle. These properties included subclinical contractures in which muscle fibers are surrounded and eventually replaced by connective tissue. These investigators also suggest that measures of threshold or a threshold-dependent variable are most indicative of spasticity. These findings would also lead us back to the more mechanical and less neurological explanations of how casting might work, as offered by Gossman et al.[2]

The use of casts in treating both contracture and motor disorders associated with hypertonicity is well established in rehabilitation practice. Practitioners have begun to identify those patients who are most likely to benefit from cast intervention.[15,19,30] These practitioners also discuss considerations regarding the timing of cast intervention and treatment that can be used to maximize and maintain the gains made with casts as well as to translate these gains into function. Regardless of the mechanisms underlying the effectiveness of cast intervention, there are differences in the casting process, types of casts, and positioning in the cast when casting a patient with significant hypertonicity, compared with casting one with primarily a muscle or soft tissue contracture. These differences are discussed in Chapters 3 and 6.

REFERENCES

1. Williams PE, Goldspink G. Changes in sarcomere length and physiological properties in immobilized muscles. *J Anat.* 1978; 127:459–468.
2. Gossman MR, Sahrmann SA, Rose SJ. Review of length associated changes in muscle. *Phys Ther.* 1982;62:1799–1808.
3. Lavigne AB, Watkins PR. Preliminary results of immobilization-induced stiffness of monkey knee joints and posterior capsule. In: *Perspectives of Biomedical Engineering: Proceedings of a Symposium of a Biological Engineering Society*, University of Strathclyde, Glasgow, June 1972. Baltimore: University Park Press; 1972;177–179.
4. Woo S, Matthews JV, Akeson WH, Amiel D, Covery FR. Connective tissue response to immobility: correlative study of biomechanical and biochemical measurements of normal and immobilized rat knees. *Arthritis Rheum.* 1975;18:257–264.
5. Noyes FR. Functional properties of knee ligaments and alterations induced by immobilization. A correlative biomechanical and histological study in primates. *Clin Orthop Relat Res.* 1977;123:210–242.
6. Kottke FJ, Pauley DL, Ptak RA. The rationale for prolonged stretching for correction of shortening of connective tissue. *Arch Phys Med Rehabil.* 1966;47:345–352.
7. Bell JA. Plaster casting for remodeling of soft tissue, part II. *The Star.* 1985;44:10–14.
8. Warren CG, Lehmann JF, Koblanski JN. Elongation of rat tail tendon: effect of load and temperature. *Arch Phys Med Rehabil.* 1971;52:465–474.
9. Warren CG, Lehmann JF, Koblanski JN. Heat and stretch procedures: an evaluation using rat tail tendons. *Arch Phys Med Rehabil.* 1976;57:122–126.
10. Goldspink G, Tabary C, Tabary JC, Tardieu C, Tardieu G. Effects of denervation on the adaptation of sarcomere number and muscle extensibility to the functional length of the muscle. *J Physiol.* 1974;270:733–741.
11. Heut de la Tour E, Tardieu C, Tabary JC, Tabary C. Decrease of muscle extensibility and reduction of sarcomere number in soleus muscle following a local injection of tetanus toxin. *J Neurol Sci.* 1979;40:123–131.
12. Tabary JC, Tabary C, Tardieu C, Tardieu G, Goldspink G. Physiological and structural changes in the cat soleus due to immobilization at different lengths by plaster casts. *J Physiol.* 1972;224:231–244.
13. Baker JH, Matsumoto DE. Adaptation of skeletal muscle to immobilization in a shortened position. *Muscle Nerve.* 1988; 11:231–244.
14. Garland D, Doyle M, Booth BJ. Early management of spastic deformities. In: Professional Staff Association of Ranchos Los Amigos Hospital, ed. *The Rehabilitation of the Head Injured Adult.* Downey, CA: Professional Staff Association of Ranchos Los Amigos Hospital, Inc; 1979:45–48.
15. Booth BJ, Doyle M, Montgomery J. Serial casting for the management of spasticity in the head-injured adult. *Phys Ther.* 1983;63:1960–1966.
16. Hill J. Management of abnormal tone through casting and orthotics. In: Kovich K, Bermann D, eds. *Head Injury: A Guide to Functional Outcomes in Occupational Therapy.* Gaithersburg, MD: Aspen Publishers; 1988.
17. Sussman MD, Cusick B. Preliminary report: the role of short leg tone-reducing casts as an adjunct to physical therapy of patients with cerebral palsy. *Johns Hopkins Med J.* 1979;145: 112–114.
18. Yasukawa A, Hill J. Casting to improve upper extremity function. In: Boehme R, ed. *Improving Upper Body Control: An Approach to Assessment and Treatment of Tonal Dysfunction.* Tucson, AZ: Therapy Skill Builders; 1988.
19. Yasukawa A. Case report: upper extremity casting: adjunct treatment for a child with cerebral palsy hemiplegia. *Am J Occup Ther.* 1990;44:840–846.
20. Yasukawa A. Upper extremity casting: adjunct treatment for a child with cerebral palsy. In: Case-Smith J, Pehoski C, eds. *Development of Hand Skills in the Child.* Rockville, MD: American Occupational Therapy Association, Inc; 1992.
21. Zablotny C, Andric MF, Gowland C. Serial casting: clinical applications for the head injured patient. *J Head Trauma Rehabil.* 1987;2(2):46–52.
22. Brennan J. Response to stretch of hypertonic muscle groups in hemiplegia. *Br Med J.* 1959;1:1504–1507.
23. King T. Plaster splinting as a means of reducing elbow flexor spasticity. *J Occup Ther.* 1982;36:671–673.
24. Maier A, Eldred E. Adaptations to long-term stretch in the passive discharge of muscle spindles. *Exp Neurol.* 1976;52:49–57.
25. Robinson KL, McComas AJ, Belanger AY. Control of soleus motoneuron excitability during muscle stretch in man. *J Neurol Neurosurg Psychiatry.* 1982;45:699–704.
26. Bessou P, Joffroy M, Montoya R, Pages B. Effect of triceps surae stretch by ankle flexion on intact afferents and efferents of gastrocnemius in the decerebrate cat. *J Physiol.* 1984;346: 73–91.
27. Carew RJ, Ghez C. In: Kandel ER, Schwartz JH, eds. *Principles of Neural Science.* New York: Elsevier; 1985:443–456.
28. Burke D, Andrews C, Ashby P. Autogenic effects of static muscle stretch in spastic man. *Arch Neurol.* 1971;25:367–372.
29. Lee WA, Boughton A, Rymer WZ. Absence of stretch reflex gain enhancement in voluntarily activated spastic muscle. *Exp Neurol.* 1987;98:317–335.
30. Hill JP. The effects of casting on upper extremity motor disorders after brain injury. *Am J Occup Ther.* 1994;48:219–224.

CHAPTER 2

Precautions and Competency in Cast Application

Paula Goga-Eppenstein, Judy P. Hill, Terry Murphy Seifert, and Audrey M. Yasukawa

PRECAUTIONS

An upper or lower extremity cast is constructed to support the limb and meet the needs and changes in the patient's extremity. The cast should be positioned properly to stabilize the force of the involved joint into optimal alignment. The joint should be positioned at submaximal range for patient tolerance. Prior to casting, several factors should be considered. For instance, the patient's skin integrity and sensitivity should be assessed. A patient with sensitive skin may require a plaster sensitivity test prior to application. To perform the test, place a small wet strip of plaster on the skin and observe the skin for an adverse reaction. Be aware of whether the patient is allergic to cotton or synthetic material and choose the stockinette material accordingly.

Minor abrasions or potential areas of skin breakdown must also be assessed and documented prior to casting. Even with these, the joint can be cast but extra care must be taken with padding and during the casting procedure to avoid increased pressure over problem areas.

A patient's ability to tolerate the procedure must also be considered prior to casting. Is the patient able to understand the reason for casts and cooperate once the casts are applied? Is the patient able to tolerate immobilization of the joint with no adverse behavior? Care must be taken when applying casts to agitated or self-abusive patients as the risk for injury increases.

Prior to casting bilateral upper extremities, be sure that medical staff members do not require access to them for monitoring blood pressure and blood drawing. For more information about specific application of casts, refer to Chapters 5 and 7.

Immediately after applying the cast, the therapist should firmly hold the desired position until the cast is set. The plaster cast typically hardens in 7 to 10 minutes. The plaster cast will dry completely in 24 to 48 hours. The patient should not bear weight on a plaster cast within the first 24 hours as this may cause an indentation and create a pressure point. The fiberglass cast hardens in 3 to 4 minutes and dries completely in 30 minutes to 1 hour. The drying time of both plaster and fiberglass is dependent on the temperature of water used during application. (Follow material manufacturer's instructions about water temperature. In general, the warmer the water, the faster the material sets. If the water is too hot and the heat becomes trapped, for example, by the cast resting on a plastic pillow, burns may result.)

After a cast is applied, the patient's caregiver must be given information about it and receive instructions on areas to be monitored. Compromise of circulation in the extremity is a common problem. Warning signs are poor nailbed refill or skin blanching postpressure; fingers and toes swollen and/or discolored; poor or absent distal pulse, dusky veins, pain, or paresthesia. If these problems occur, removal of the cast is recommended to avoid skin breakdown, deep vein thrombosis, and other difficulties. If skin integrity is thought to be compromised, there may be reddened areas just proximal and/or distal to the cast or a patient may complain of rubbing or pressure on the extremity. If the patient describes a pain that is throbbing, specific to the muscles being stretched, and it does not resolve within 4 hours of cast application, the joint has been overstretched and casted in maximal range. At that time, the cast should be removed and replaced with the joint positioned at submaximal range.

To be sure that patients and caregivers are aware of warning signs, written instructions should be utilized. For the inpatient, document in the chart which extremity and joint have been casted and the areas to be monitored. Provide written instructions for outpatients and advise them that if the cast must be removed immediately they should go to the nearest emergency room (ER). Provide the patient and the caregiver with a letter to present to the ER physician describing the purpose of the cast, that it was applied by a therapist, and that it can be removed for emergencies. Record on the cast the date of application and estimated removal date.

In addition to precautions and emergency procedures, the patient and caregiver should be instructed in basic comfort and hygiene as it relates to the cast. When bathing, care must be taken to avoid getting the cast wet. Instruct the patient and caregiver to wrap the extremity in a plastic bag and tape the bag securely at the proximal end. If the plaster cast becomes wet, it softens and loses its shape. Moisture inside the stockinette and padding may cause skin irritation or breakdown. Advise patients that if itching occurs under the cast, they should not scratch their skin using a stick or hanger. This could result in severe skin breakdown.

To improve the patient's tolerance and comfort while wearing the cast, instruct the patient to wear loose-fitting clothing to fit over the cast. When they are sleeping, suggest that patients place pillows around the cast to provide cushioning.

The serial cast is left in place for 5 to 7 days. After removal of the cast, the limb may look dry with flaky skin. Clean the skin with warm, soapy water. Check the skin for red areas, blisters, and breakdown. The joint should then be stretched and mobilized to increase the lubrication to the joint and prevent adhesions from forming. Range of motion is then assessed and gains documented. If required, the joint is then repositioned at the new submaximal range.

During the casting process and at its completion, an individualized therapy and home exercise program should be established by the therapist to ensure continued progress. A successful casting program is a team effort among the patient, the caregivers, and the therapists. Education and continuous monitoring of the patient's progress are the keys to a problem-free program.

ENSURING COMPETENCY WITH CAST APPLICATION

Casting is an advanced treatment technique that carries some risk of skin breakdown and vascular and sensory complications if applied incorrectly. Competent use of the technique requires knowledge of anatomy, neurologically impaired movement, and functional assessment as well as skill in cast application technique. To ensure quality service delivery a therapist must integrate casting knowledge and skills while subscribing to a professional standard of care. See Appendixes A and B for policies and procedures about casting of the upper and lower extremities from which a standard of care and monitoring system can be derived. A therapist's clinical knowledge base and ability to perform specific casting techniques affect the cost effectiveness and outcome of the procedure and also the patient's and family's understanding and perception of their roles related to the casting protocol.

Examples of competency efforts include measuring, assessing, and providing feedback to improve the performance of the clinical staff. These efforts must be carried out by an experienced therapist. The methods commonly used in the process include

- education, such as attending a casting workshop
- practice of techniques learned at a workshop with peer feedback
- evaluation of casting technique
- assessment of clinical knowledge base, such as regarding contracture and spasticity
- continuous monitoring of cast effectiveness and complications
- annual reassessment of staff competency

The competency checkout used for casting may include an oral review to measure an individual's knowledge base and clinical reasoning, a demonstration of technical skills, and a demonstration of performance level. Competency can be demonstrated by effective determination of when to cast and what type of cast to use, technical skill in cast application, assessment of results, and determination of appropriate techniques to use in conjunction with and following casting. Both clinical reasoning and technical skill are necessary for competent cast application. Appendixes C and D provide examples of the guidelines for the checkout process.

Forms for recording the data used for monitoring quality assurance efforts and patient care outcomes are included in Appendix E. This worksheet serves as a basis for analyzing the outcome and effectiveness of casting. It also gives the therapist access to baseline and outcome data on each patient from which outcome and effectiveness of the casting program can be analyzed. This form summarizes findings and cannot replace the comprehensive diagnostic assessments done initially and upon discharge and documented in the individual patient medical record.

The upper and lower extremity casting quality monitor log shown in Appendix F is an example of a specific monitor that can be used to track complications resulting from casts. For example, a quality assurance study can be performed using one of the logs to identify the number of incidents of skin lacerations that occurred upon removal of casts. The data collected can be analyzed so that the problem can be clearly identified and actions taken to ensure that proper technique is used in the future.

Education of the patient and caregivers is essential before the start of a casting program. The patient's and family's understanding of the time commitment needed for attending ongoing scheduled sessions and following through at home is important to achieving a positive outcome. To confirm that the patient and caregiver were given an explanation of the rationale and a description of the casting procedure, a consent form is often used. See Appendix G. The caregiver who is taught cast care adequately and understands the rationale of the program is more likely to continue with the program at home.

Appendix H is a form given to outpatients or their caregivers that reminds them of the scheduled date of return as well as precautions to monitor. A letter addressed to the ER physician, as shown in Appendix I, will provide some of the information he or she needs to feel comfortable removing the cast in emergency situations.

CONCLUSION

These sample forms and protocols can be used to develop specific procedures and a monitoring system to reflect your facility's policies. Specific casting protocols will need to be updated periodically. Protocols are useful for developing standards of quality patient care, especially for complex diagnoses such as juvenile rheumatoid arthritis and traumatic head injury.

Instructional sheets can be given to the patient and caregivers that can augment their follow-through at home. However, they cannot take the place of the therapist as a teacher. The therapist should use words that are simple and easily understood and that reinforce the instructional sheets. Good patient and caregiver teaching helps ensure follow-through with the program and encourages the family to be supportive. This is why the therapist must work with the patient and educate and involve the family as well.

Competent quality practice comes with the experience and clinical judgment of the therapist. Therefore, casting competency can only be developed as a result of hands-on experience provided under the guidance and supervision of a clinician who is highly experienced in the technique and has developed sound clinical judgment over time.

CHAPTER 3

The Use of Casts To Manage Upper Extremity Motor Disorders

Judy P. Hill and Audrey M. Yasukawa

CASTS TO MANAGE CONTRACTURES

Contracture management casts incorporate positions to put contracted muscles or soft tissue near the end of their elastic limits. Left immobilized in this position, cell division and increase of sarcomeres can occur as discussed in Chapter 1. With most upper extremity muscles crossing multiple joints, the position of maximal muscle elongation must be considered. For example, the biceps brachii acts as an elbow flexor, forearm supinator, and shoulder flexor and internal rotator. When attempting to correct a contracture in the biceps, positioning the elbow in increasing degrees of extension may cause the forearm to supinate, especially if the pronators are weak or not functioning. In this case, correcting the contracture may require a cast that can provide elbow extension and forearm pronation simultaneously. Similarly, if the flexor digitorum profundus is contracted, extension at the wrist will cause secondary flexion of the digits. Extension of the fingers will cause secondary flexion at the wrist. A cast that can provide finger and wrist extension will be necessary to manage the contracture. Casts commonly used to manage contractures are described below. Chapter 5 includes more specific descriptions of each cast, the rationale for its use, and fabrication instructions.

Elbow Drop-Out Cast

The elbow drop-out cast is frequently used to manage severe elbow flexion contractures. With this cast

preventing flexion of the elbow, gravity can assist in pulling the forearm into extension away from the humerus (Figure 3–1). Two types of drop-out casts exist: the forearm enclosed cast (Figure 3–2) and the humeral

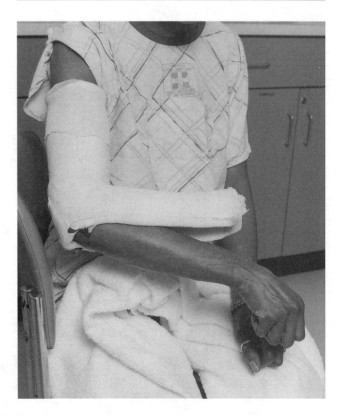

Figure 3–1 Elbow drop-out cast with humeral portion enclosed.

9

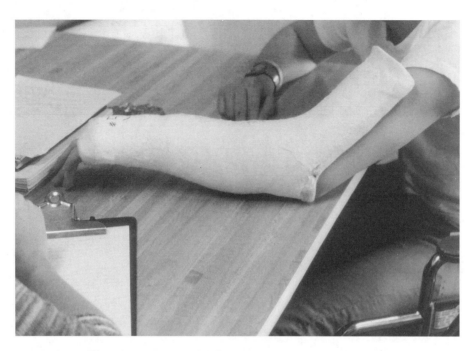

Figure 3–2 Elbow drop-out cast with forearm enclosed.

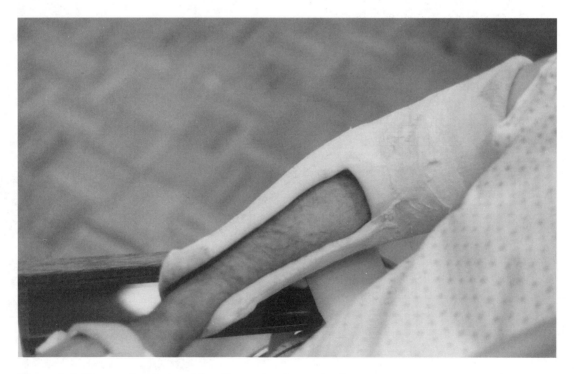

Figure 3–3 Reverse drop-out cast for elbow extension contractures.

portion enclosed cast (Figure 3–1). The reverse drop-out cast is used for elbow extension contractures (Figure 3–3). All of the drop-out casts are most effective when the limb can be placed in a position for gravity to have an effect on the forearm by pulling it into the desired direction. This usually requires the patient to be upright, sitting or standing, for the majority of daytime hours to position the limb in a way that will allow gravity to pull the forearm into extension. When elbow flexion contractures are less severe, a drop-out cast may slip down on the forearm as the limb extends. For these less severe contractures (45° flexion or less), the rigid circular elbow cast (Figure 3–4) is used.

Rigid Circular Elbow Cast

The rigid circular elbow cast does not have the advantages of the drop-out casts of allowing the range gain to be observed with the cast in place and gravity cannot assist in extending the limb. The rigid circular elbow cast, however, does maintain the muscles in an extended position and avoids the cast slipping that can occur with drop-out casts once the elbow is extended beyond 40° of flexion.

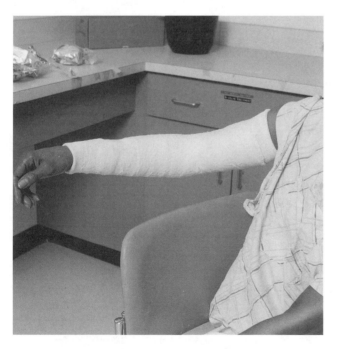

Figure 3–4 Rigid circular elbow cast.

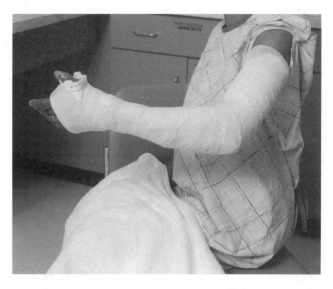

Figure 3–5 Long arm, elbow-forearm-wrist cast.

When elbow flexion contracture is present and forearm pronators are weak or denervated, the cast must control biceps length across the elbow and forearm joints. This is also necessary when the biceps is severely contracted, causing elbow flexion and supination contractures. The most effective cast in these cases is the long arm, elbow-forearm-wrist cast (Figure 3–5). Attempts to utilize a rigid circular elbow cast and control the forearm position by conforming the cast to the distal radial-ulnar forearm joint are usually not successful. Often the forearm rotates in the cast even with the forearm portion conformed to the oval shape of the distal forearm (Figure 3–6). Also, risk of excessive pressure exists, with the forearm pulling into supination as a result of the pull on the biceps across the elbow. With the cast extended down over the wrist and using the metacarpals for leverage, the pressure for controlling forearm position is distributed over a larger area, making excessive pressure less likely. The drop-out casts will not be effective in these cases as they allow the forearm to rotate. As the elbow extends, the forearm may be pulled into increasing supination.

Wrist Cast Combined with Finger Cast

Extrinsic hand musculature contractures will require casting of the wrist and finger joints due to the muscles crossing the wrist and at least one joint in the hand.

Figure 3–6 Rigid circular elbow cast with forearm conformed to shape of the distal forearm.

Figure 3–7 Wrist cast with multiple finger casts.

Management of these contractures can be accomplished with the wrist cast combined with individual or multiple finger casts (Figure 3–7) or a finger shell encompassing all of the fingers (Figure 3–8). With these casts, the musculature can be serially extended by changing the position of either the wrist or finger joints. For example, the finger shell can be made with the proximal interphalangeal (PIP) and distal interphalangeal (DIP) joints extended. Because it is attached to the wrist cast with hook and loop fasteners, the metacarpophalangeal (MCP) joints can be allowed to remain somewhat flexed (Figure 3–9). As elongation of the muscles occurs, the hook and loop can be tightened or shortened, pulling the MCPs into further extension

(Figure 3–10), which further stretches and elongates the finger flexors. Alternatively, the finger shell can be fabricated in more PIP flexion and attached to the cast in more MCPs extension. Another option would be to fabricate the finger shell with the PIPs in extension, attach the hook and loop with MCPs in extension, and allow more flexion in the wrist cast. In this case, as elongation occurs, the wrist is recast in more extension. For thumb extrinsic muscle contracture, similar procedures are used, with the thumb fully enclosed and incorporated with a wrist cast.

Also, contractures exist that appear to be isolated to the hand but actually occur secondary to wrist contractures as a result of extrinsic hand musculature also crossing the wrist. For example, the hand may be in an intrinsic minus position with the MCP joints in hyperextension and the PIPs flexed. If the wrist is contracted in flexion and the finger flexors are weak or denervated, the extensor digitorum communis, being elongated across the wrist, may be causing the intrinsic minus position

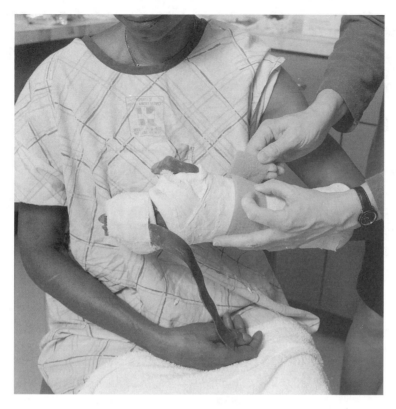

Figure 3–8 Wrist cast with finger shell encompassing all fingers.

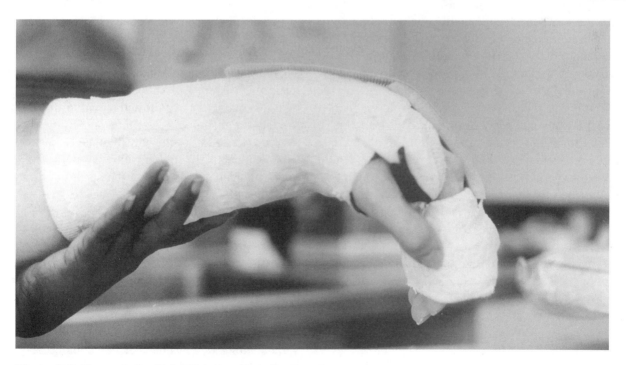

Figure 3–9 Finger shell with MCP joints somewhat flexed.

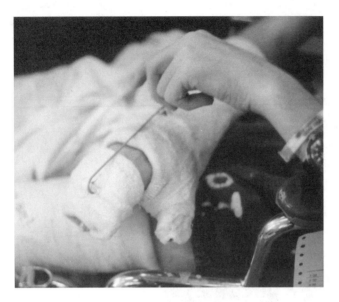

Figure 3–10 Finger shell with MCP joints in further extension.

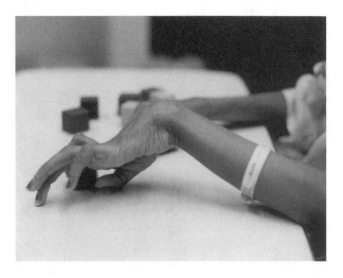

Figure 3–11 The wrist is contracted in flexion and the finger flexors are weak or denervated, causing an intrinsic minus position.

(Figure 3–11). In this case casting the wrist into extension may allow enough slack on the finger extensors to correct the apparent intrinsic minus position in the hand without finger casting (Figure 3–12).

Contractures of the intrinsic muscles of the hand will require consideration for the multiple joints they may cross. The lumbricales, if shortened, will cause MCP flexion contracture. When attempting to correct the contracture by casting the MCPs in gradually increasing extension, PIP flexion may need to be addressed as well. This is due to the attachment of these muscles and their role in simultaneously flexing the MCPs and extending the PIPs. As contracted lumbricales are stretched across the MCP joints, they may cause extension, even hyperextension, of the PIPs if these joints are not casted into flexion or managed in another way to pull them into flexion. Finger flexion loops can be used in conjunction with the cast for this purpose (Figure 3–13).

Prolonged positioning of the fingers in a deforming position can result in contracture and adhesion in the soft tissues surrounding the joints and even the skin. Because the joints are small, significant secondary contractures can occur rather quickly. Individual finger casts are quite effective in correcting these contractures (Figure 3–14).

Contractures in the burned patient result from damage to the cutaneous layers and sometimes muscle. Usually significant scar formation and loss of tissue elasticity occur. Casting has been effective with managing these types of contractures as well as in providing constant pressure to reduce scar formation. Most commonly drop-out and rigid circular elbow casts, wrist casts, and individual finger casts are used for localized elongation. A tendency exists for the gains made with casting to be difficult to maintain with these types of contractures, as the active scar formation tends to result in resumed shortening once the cast is removed. The ability of the patient to use opposing musculature to actively stretch following the casting program is crucial in maintaining the results.

Summary

Casting to manage contracture takes into consideration the etiology of the contracture and the entire muscle or soft tissue length as they cross multiple joints. The basic approach is to provide low-load, long-duration elongation of the tissues to allow sarcomere division and cell division to occur. To ensure that the optimal elongation has been achieved in the cast, joint position across all joints that the muscle crosses must

Figure 3–12 Casting the wrist into extension to correct the intrinsic minus position.

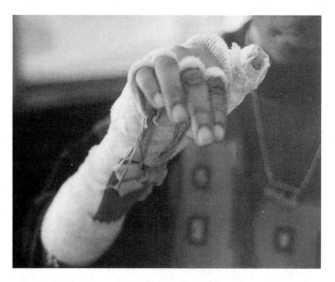

Figure 3–13 Wrist cast finger flexion loops.

be considered. Both case report and small subject pool–controlled research articles have been published that support the effectiveness of casting in altering soft tissues and muscle length as discussed and referenced in Chapter 1.

CASTS TO MANAGE ABNORMAL MUSCLE TONE AND MOVEMENT IN THE UPPER EXTREMITIES

Spasticity is a major problem interfering with motor function. Individuals with central nervous system damage may have abnormal tonic stretch reflexes and involuntary muscle contracture or spasms. This may produce abnormal muscle tone, which may be seen clinically as spasticity. The muscles that tend to develop spasticity are generally shortened and in a hypertonic state. There is increased resistance to passive stretch due to the increased responses of static and often phasic stretch reflexes.[1] The hyperactive stretch reflexes have been cited as the primary problem underlying loss of volitional control. Lee et al.[2] suggested that changes in reflex threshold angle characterizes spastic

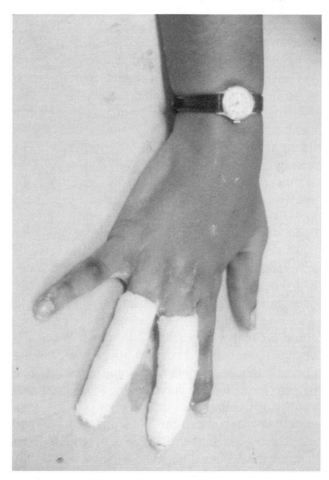

Figure 3–14 Individual finger casts to correct contracture.

limbs better than changes in joint stiffness. Sahrmann and Norton[3] examined spasticity and abnormal movement as two separate factors to consider in upper motor neuron syndrome. They suggested that the hyperactive stretch reflex is not the limiting impairment of movement, but rather the inability of the agonist to contract appropriately during volitional movement.

Abnormal tone may be the direct result of a lesion of supraspinal, spinal, and/or congenital origin. This may produce varying degrees of hypertonicity and restriction in upper extremity movement. In the upper extremity, the usual distribution of hypertonicity appears in the antigravity muscles. The upper extremity musculature is positioned in a flexion attitude, accompanied by abduction with internal rotation at the humerus, pronation at the forearm, flexion at the elbow, and flexion at the wrist and fingers, with an adducted or indwelling thumb. The shortened muscles may account for the contracture and muscle imbalance. This muscle imbalance may be due to a combination of hypertonicity and paretic muscle weakness. Any voluntary movements may be masked partially or totally.

The arm movements, when present, are often stereotyped in a synergistic or compensatory pattern. The movements may be associated with cocontraction of the antagonist muscles. The agonist, in this case the flexor musculature, is much more powerful. This muscle imbalance may be exacerbated by a contracted or shortened agonist and overstretched antagonists that interfere with the production of isolated, selective movements of the arm. If isolated movement is present, often it is slow, stiff, and uncoordinated. Upon active movement initiation, it may be difficult to relax the arm. Usually the compensatory arm movement results in fixed, abnormal posturing, which leads to subsequent muscle shortening, contracture, and joint tightness. The functional consequences for the patient include difficulty in performing daily living skills. Despite therapeutic intervention of daily range of motion, splinting, and various therapeutic techniques, contracture formation remains a major sequela for individuals with abnormal muscle tone. The rationale underlying upper extremity casting is to prevent contractures and further limitations of motion by decreasing the abnormal muscle contraction responsible for the muscle shortening. Upper extremity "inhibitory" casting has been used in the prevention and treatment of contractures induced by abnormal muscle tone and movement. The implication of inhibitory casting is that it reduces the abnormal tone to facilitate more normal movement patterns. Several research studies and single case reports investigated the use of casting as an effective treatment method in the reduction of spasticity and facilitation of active motor patterns.[4-7]

The types of casts used to incorporate inhibition of abnormal tone in the upper extremities are the same as those used for contracture management as discussed in the first section of this chapter. The casts are applied incorporating positions that have been found to help decrease tone in the extremities. For example, if thumb abduction and extension are found to have a relaxing effect on the fingers, the thumb can be placed in this position in the cast (Figure 3–15), while the patient works on active finger extension with the cast in place. Likewise, if supinating the forearm has a relaxing effect on elbow extension, supination can be incorporated in a long arm cast (Figure 3–5) to foster more extension and elongation of the arm.

In some cases positioning the elbow in extension and relaxing the flexor musculature has a relaxing effect distally at the wrist and into the hand as well. A rigid circular elbow cast (Figure 3–4) can be used to take advantage of this relaxation effect. In other cases, when the abnormal tone is affecting the wrist and hand more than the elbow, a rigid circular wrist cast, often with the thumb enclosed as well, can be used to inhibit. The relaxation of the wrist and hand along with the weight of the cast helps to foster relaxation and extension of the elbow.

When an individual exhibits moderate to severe finger and/or wrist flexor tone in the absence of active extension, the finger shell (Figure 3–8) can be added to the rigid circular wrist cast. In cases with very mild finger flexor tone that interferes with full simultaneous wrist, finger, and thumb extension, the platform cast (Figure 3–16) can be used. This cast often results in improved manipulation skills as the mild tone is inhibited.

The method by which casting, with its slow, gradual stretch and static positioning, might inhibit abnormal tone is somewhat uncertain. Clinical observation suggests that reduction of hypertonicity does not always result in a simultaneous improvement in function and control. Brennan[8] examined 14 patients with hemiplegia of cerebral-vascular origin. He compared changes

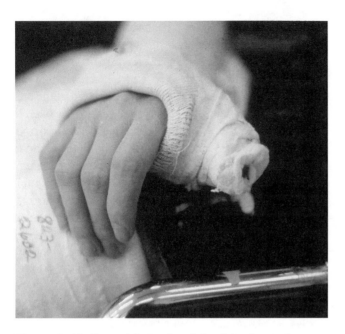

Figure 3–15 Cast with the thumb in abduction and extension.

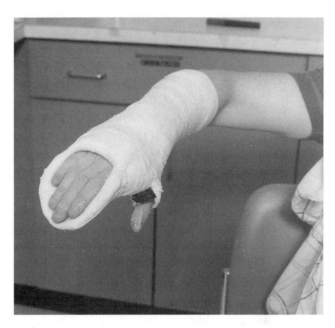

Figure 3–16 The platform cast.

of the stretched, hypertonic flexor muscle group and the antagonist extensor group with muscles in a corresponding neighboring joint that were left untreated. The upper limb posture was predominantly of the flexor muscle group with constant tonic spasm of the affected side. The treatment consisted of wearing a splint on the joint held in flexion by keeping that joint extended in a splint. The splint was worn for about 3 months and removed only for hygiene purposes once a day. He found that through the prolonged periods of stretch, in most cases, there was a decrease in hypertonicity and an increase in active opposing movements, which either improved function or the limb posture. In most cases there was an increase in strength and active movement of the antagonist and improvement in range.

Mills[9] compared electromyographic (EMG) activity of spastic muscles in eight clients with brain injury before and during splint application. He found an increase in joint position into extension in the splinted condition with elongation of the spastic muscles versus with the nonsplinted condition. In this study, there was no statistical difference found in the EMG activity. This might lead to the assumption that splinting can control the abnormal posturing caused by hypertonicity without actually altering the muscle tone, as measured by

EMG. The lack of movement in the antagonist may be as much the problem as hypertonicity in the spastic limb. These studies may suggest that the resistance to passive movement seen in patients with central nervous system dysfunction may be associated with reduction in muscle length as well as the reduced potential for the muscle to activate. The use of casting may assist with altering muscle length, improving the balance of the agonist and antagonist, and—in some cases where adequate volitional control is present—result in more normal movement patterns.

Several centers treating patients with traumatic brain injury routinely use serial application of plaster casts for improving passive and/or active range of motion of the upper extremity.[10,11] Throughout the casting program the abnormal compensatory movements are inhibited and the desired motor control, rebalancing of agonist and antagonist, and joint alignment are progressively facilitated with gradual stretch and relaxation.

Individuals with persistent hypertonicity and limited opposing muscle activity may require passive lengthening with a series of casts to stretch the tight, spastic muscle gradually to the desired position. The primary goal of casting in these cases is to prevent contractures and maintain available range of motion to enable a caregiver or the individual to perform daily care and hygiene and to position the extremity properly. Single case reports document the use of casting to effectively manage abnormal tone and maintain passive range of motion.[12–14]

Summary

The contour of the upper extremity cast may provide the inhibitory components of deep pressure, neutral warmth, and prolonged stretch as well as interrupt stretch input to the muscle spindle that could fuel the hypertonic response. With the cast in place, no changes in muscle length can occur. It is these changes in muscle length that cause excitation of the muscle spindle afferents, which then cause excitation of the extrafusal muscle fibers via the anterior horn of the spinal cord. The type of upper extremity cast to be used to produce the relaxation and decrease the abnormal muscle tone or movement is individually determined to incorporate the position that best results in the desired relaxation for a particular individual. Rigid circular casts are more commonly used to manage abnormal tone. Drop-out casts are seldom used unless significant contracture has already occurred.

REFERENCES

1. Bishop B. Spasticity: its physiology and management, parts I–IV. *Phys Ther.* 1977;57:371–401.
2. Lee WA, Boughton A, Rymer WZ. Absence of stretch reflex gain enhancement in voluntarily activated spastic muscle. *Exp Neurol.* 1987;98:317–335.
3. Sahrmann SA, Norton BJ. The relationship of voluntary movement to spasticity in the upper motor neuron syndrome. *Ann Neurol.* 1977;2:460–465.
4. Hill JP. The effects of casting on upper extremity motor disorders after brain surgery. *Am J Occup Ther.* 1994;48:219–224.
5. Law M, Cadman D, Resenbaum P, Walter S, Russell D, DeMatteo C. Neurodevelopmental therapy and upper extremity inhibitive casting for children with cerebral palsy. *Dev Med Child Neurol.* 1991;33:379–387.
6. Tona JL, Schneck CM. The efficacy of upper extremity inhibitive casting: a single-subject pilot study. *Am J Occup Ther.* 1993;47:901–910.
7. Yasukawa A. Case report: upper extremity casting: adjunct treatment for a child with cerebral palsy hemiplegia. *Am J Occup Ther.* 1990;44:840–846.
8. Brennan J. Response to stretch of hypertonic muscle groups in hemiplegia. *Br Med J.* 1959;1:1504–1507.
9. Mills VM. Electromyographic results of inhibitory splinting. *Phys Ther.* 1984;64:190–193.
10. Booth BJ, Doyle M, Montgomery J. Serial casting for the management of spasticity in the head-injured adult. *Phys Ther.* 1983;63:1960–1966.
11. Hill J. Management of abnormal tone through casting and orthotics. In: Kovich K, Bermann D, eds. *Head Injury: A Guide to Functional Outcomes in Occupational Therapy.* Gaithersburg, MD: Aspen Publishers; 1988.
12. Cruikshank DA, O'Neill DL. Upper extremity inhibitive casting in a boy with spastic quadriplegia. *Am J Occup Ther.* 1990;44:552–555.
13. King T. Plaster splinting as a means of reducing elbow flexor spasticity. *Am J Occup Ther.* 1982;36:671–673.
14. Smith LH, Harris SR. Upper extremity inhibitive casting for a child with cerebral palsy. *Phys Occup Ther.* 1985;5:71–79.

CHAPTER 4

Upper Extremity Assessment

Judy P. Hill and Audrey M. Yasukawa

Assessment used with upper extremity casting includes assessing for the efficacy of cast intervention, comprehensive pre- and postcast intervention assessment, and targeted assessment between casts in a series. Components in the assessment include passive range of motion, muscle length, spasticity, active motion, functional use, sensation, position of the limb at rest, and postural influences on limb position.

DECIDING TO CAST

The primary clinical manifestations that might suggest considering cast intervention are passive limitation of motion, limitation in active motion and movement impeded by postural influences, and increased muscle tone. When these are present, further assessment to determine whether casting might be indicated is initiated. This includes determining whether there might be any orthopaedic impairments, usually by discussion with the physician who has reviewed x-rays. In general, losses in range of motion due to bony impairments are not altered by casting. In the case of heterotopic ossification, casting is usually contraindicated because mobilization of the joint(s) affected by the ossification is recommended and cannot be done with the cast in place. Occasionally, the probable benefits of casting and increasing range in one direction outweigh the risk of increased stiffness from the heterotopic ossification, and casting is initiated with the approval of the managing physician and/or orthopaedic specialist.

Another factor to help determine whether casting might be beneficial is review of the patient's response to treatment to date. A patient who has not been able to participate in active therapy may be given a trial of more active treatment interventions and mobilization prior to initiating casting to see whether these methods alone might result in improved motion. A patient who has had splinting, passive range of motion, and some opportunity to participate in functional activities but whose upper extremity remains contracted and has no x-ray findings might be considered immediately for casting. A patient who has shown some motor recovery and functional use but seems to have plateaued, still demonstrating imbalanced and patterned motion as well as some contracture, might also be considered immediately for casting to balance the muscle activity and improve functional use. A patient with similar motor skills but who has a significant disregard for the extremity, however, may benefit more from an active facilitation program than from casting.

Assessment for functional gains expected and patient/family goals for casting are also important in determining whether casting might be indicated. A patient may have a range of motion limitation along with lack of any movement or ability to use the extremity in functional tasks. Casting might be effective in increasing range of motion, but has no real benefit or value for the patient with no active motion and impaired sensation. Casting to increase range of motion in this situation could have value for some patients if their goal is

19

to improve the cosmetic appearance of the limb, to make the extremity easier for them to manage passively, or to allow adequate hygiene to prevent skin breakdown. An example of a significant cosmetic reason for casting might be posed by a patient who expresses a goal of having the arm at the side when ambulating instead of its being positioned across the chest in elbow flexion. In this case, casting might result in more elbow extension, making the patient more comfortable with the arm and possibly also making it easier for the patient to put on clothing with sleeves.

Once casting is determined to be a viable intervention for a particular client, assessment focuses on selecting the type of cast to use.

PRECAST AND POSTCAST ASSESSMENT

Once the determination is made that casting is indicated, more specific evaluations are performed to document baseline measures. These baseline measures are taken for areas that may be affected by casting. The measurements are taken before the first cast and after the last cast in a series. Individualized evaluation components may be utilized that are specific to the patient's goals. That is, if the patient is hoping to use the extremity functionally during dressing, specific observations of how the patient incorporates the extremity in dressing prior to casting are documented as part of the pre- and postcast assessment. Videotaping can be a useful method of comparing function before and after a casting program. Standard evaluation items are given below.

Passive Range of Motion

Goniometric range of motion measurements of the entire extremity to be casted should be recorded. It is possible to see changes in range, both positive and negative, in joints adjacent to the one(s) casted. This is likely due to the effect of casting on muscles that cross more than one joint. For example, with an elbow cast in place, increased range in elbow extension might be noted following cast removal. On measuring forearm pronation a decrease might be noted due to the biceps pulling the forearm into supination as it is stretched across the elbow. Passive range of motion measurements should be taken with consideration for total arc of available mo-

tion rather than range in one direction only. For example, with an elbow flexion contracture, if maximal extension prior to casting was 50° of flexion and following casting maximal extension was 10° of flexion, a gain of 40° might be claimed. If, however, prior to casting the patient had 120° of elbow flexion and following casting had only 95° of flexion, the gain in the arc of motion would be 15°. These measurements should be taken at least several hours after removing the final cast in a series to allow for time to range the joint both actively and passively, as some stiffness as the result of static positioning is expected immediately following cast removal.

Muscle Length

Goniometric measurements of joints crossed by multijoint muscles can be used to assess muscle length. In doing this, only one joint at a time should be allowed to vary. Elbow range can be measured with the forearm positioned in maximal pronation or supination, or neutral or forearm range can be measured with the elbow in a set amount of flexion. When casting to achieve increased finger extension, finger range is measured with the wrist in a fixed position, or wrist range is measured with fingers in a fixed position, usually maximal extension to get an indication of muscle length. When comparing these measurements before and after cast intervention, the "fixed" joint position must always be the same.

Muscle Contracture versus Joint Limitations

While evaluating passive range of motion and muscle length, the relation between muscle contracture and joint range limitations must be considered. If, when the wrist is flexed to its maximum, the fingers can be more fully extended than when the wrist is in maximal extension, a muscle contracture is the problem. When the wrist position (freeing the finger flexors over the wrist) holds no influence on the finger joint position, the problem is with the joint and soft tissues around the joint. This makes a difference in determining what type of cast will be most effective. When the problem is more isolated to the joint, a cast can be applied to that joint. When the problem is muscle contracture, the cast must control the muscle position across all (or as many as possible) joints, so a cast that holds the position of multiple joints simultaneously must be used.

Sensation

Sensation evaluation in the extremity to be casted is indicated for two reasons. First, sensory impairment has implications for functional use. A patient may not be expected to have as good a functional outcome with significant sensory impairment. Sensory impairment as well as motor deficits may be contributing to manipulation problems. Second, changes in sensory status might occur during or following casting if the cast is too tight or creates pressure. A baseline record of sensation is important for comparison should there be any question of negative effects of casting.

Spasticity Or Abnormal Muscle Tone

Reliably measuring spasticity is a recognized problem in rehabilitation and neurophysiology. With spasticity defined as a motor disorder characterized by a velocity-dependent increase in tonic stretch reflexes, it can be extrapolated that measuring it must include controlling for velocity, tension, and length. It is difficult to control all of these variables reliably in the clinical setting. While recognizing this, some crude indicators of spasticity can be used to compare pre- and post-casting. The Ashworth Scale classifies spasticity as mild, moderate, or severe depending on the range of the joint where the quick stretch reflex is elicited. The joint angle or muscle length at which the quick stretch reflex is elicited can also be measured with a goniometer. When taking this measure, it is important to take into account the issues covered in the muscle length assessment section. All joints crossed by the muscle in which the quick stretch is being measured must be stabilized in a fixed position. When taking a quick stretch measurement at the elbow, the forearm should be stabilized in a specific position, preferably neutral, and the position recorded for later comparison.

Motor Control

The patient's ability to perform specific movements and the speed with which they can be performed are assessed noting total limb position. The ability to flex the shoulder while maintaining elbow extension and then actively supinating and pronating the forearm is observed. The ability to bring the hand to the mouth and back to the table top while maintaining pronation is another example. To assess the speed of motion, the number of times a patient can flex and extend the elbow, supinate and pronate the forearm, and grasp and release in 10 seconds is measured.

Functional Use

While motor control assessment documents the patient's ability to perform various motions on command, functional use assessment focuses on the spontaneous incorporation of the extremity in activities of daily living. A standard group of tasks that require a variety of upper extremity motions can be used, as well as tasks devised for a particular patient.

Baseline Skin Condition and Circulation

Observations of skin surfaces should be documented. Pulses at wrist and in thumb should be located and assessed for comparison with their status with the cast in place. Any tendency for edema should be noted and measured distally, where it can be monitored with the cast in place. Temperature of the hand to be casted compared with the uncasted hand can also be noted for comparison with status once the cast is in place.

Note: While the above assessments are recommended for patients prior to initiating a casting program, there may be cases where not all of them can be performed. A patient with active movement will require all of the above assessments, while a patient with little active motion or very low cognitive functioning may have only spasticity, range of motion, and baseline skin condition and circulation assessment.

ASSESSMENT BETWEEN CASTS IN A SERIES

While attention to all of the areas above is important in assessing the effectiveness of a casting program, including a series of casts, between casts in a series the focus is on removing and replacing the cast as quickly as possible to incorporate the gains from one cast to the next. With this in mind, assessment between casts in a series focuses on ensuring that gains are being made in range of motion, spasticity reduction, and functional use and that no negative complications such as skin breakdown are noted.

ASSESSMENT BETWEEN SERIAL CASTS

Skin Condition and Circulation Assessment

Observations of skin condition should be made with attention to any red or open areas resulting from the cast. If present, red areas should blanch and resolve within 15 minutes. Areas that do not resolve within this time frame and open areas require special consideration if the casting program is to continue. A better-fitting cast or a cast with a window cut over the problem area may be considered. If there is actual skin breakdown, special dressings may be used and casted over, under the direction of the physician. In some cases, the cast program may need to be put on hold until the problem skin area resolves. Edema is also noted and, if significantly increased compared with the status before cast application, may require delaying cast reapplication or consideration of an alternative type of cast.

Range of Motion Assessment

Between casts in a series, passive range of motion of joint(s) casted is measured and compared with the measurements taken prior to applying that cast. A gain of 15° to 20° is expected. With a gain of less than 10°, continuation of the casting program might be questioned. Things to consider in this are how long-standing the contracture was and whether the cast might need to be left in place for several more days to see more gain. The cast that was removed can be bivalved and replaced on the arm to test this possibility. A second cast might be applied, but if no more gain results from that cast, the program should be discontinued and other treatments initiated. If range gain is minimal from a cast, another type of cast might also be considered. While consideration of total arc of motion, possible stiffness, and loss of motion in the opposite direction are important in assessing the results of a casting program, they are not as important between casts in a se-

ries when the focus is on getting the next cast on quickly. Some stiffness is expected after cast removal. The arm might be ranged actively and passively for approximately 15 minutes to counteract the stiffness and preserve the total arc of motion. If more than 10° of range is gained in the desired direction, the program is usually continued even if there is some loss of ability to move in the opposite direction. In most cases this stiffness can be worked out with active therapy following the casting program.

Spasticity Assessment

Goniometric measurement of joint position where the stretch reflex is initiated is taken between casts in series as an indicator of change in spasticity. This measure is usually not used as an indicator of whether to continue with the casting program. If there is no change in spasticity, but there is passive range of motion gain and no skin condition complication, the program is continued with the next cast in the series.

Functional Use Assessment

A brief functional assessment might be performed. This can include reaching and/or grasping and releasing objects or a specific task such as putting the arm into a sleeve, putting on a glove, or tying a shoe. This can encourage the patient in the results of casting and indicate to the therapist types of treatment that might be indicated with the cast in place and following the casting program to facilitate functional use.

CONCLUSION

Assessment is essential in both predicting and evaluating the effectiveness of any casting program. It provides the basis for deciding what type of cast to use. Assessment also indicates to the therapist whether the specific patient with his or her particular clinical presentation is likely to achieve desired goals through casting. Therefore, proper assessment is key to effective use of casting in treatment.

CHAPTER 5

Upper Extremity Casts: Types and Application Descriptions

Judy P. Hill and Audrey M. Yasukawa

RIGID CIRCULAR ELBOW CAST

Rationale for Use

A gradual loss of elbow extension with myostatic contracture may result in a serious impairment and interfere with a patient's performance of activities of daily living. The rigid circular elbow cast can be applied in a series to improve range of motion. The elbow cast is effective for improving extension in patients with mild spasticity, or those treated with a series of drop-out casts who now present with less than a 45° limitation. In patients with fluctuating tone, the elbow cast may assist with providing equalized pressure for gradually improving range. If the individual presents with severe rigidity or heterotopic ossification and mobility is required, the elbow cast should not be used.

For those with severe to moderate spasticity, the elbow cast may be preferred to the drop-out cast because of the ease of application. However, it is important when casting the elbow to apply equalized pressure throughout the humerus and forearm, to provide an effective counterpoint, and to prevent pressure to the olecranon. The padding for the elbow cast can create varying amounts of pressure, depending on how it is wrapped. The olecranon is wrapped in a figure-eight, rather than a circular fashion, to avoid direct pull and tension or tissue constriction on this bony prominence. If a circular wrap is used, pressure can be uneven and

be placed directly on the olecranon. The figure-eight padding wrap helps to distribute the pressure evenly.

The rigid circular elbow cast is left in place for 5 to 7 days. After its removal, the arm should be checked and cleaned, and the range of motion documented. A new cast can then be applied immediately in a few degrees more of extension depending on the direction casted.

Required Casting Materials

(See Appendix J for a listing of casting supplies and equipment.)

- Stockinette: 3-inch width for most adults; 2-inch width for very small adults and children
- Cast padding: 3-inch width, three to five rolls depending on size of arm
- Plaster: 3-inch width, three to five rolls depending on size of arm
- Felt: four strips
    one 1½-inch width by circumference of humerus just below axilla
    one 1½-inch width by circumference of forearm over ulnar styloid
    two 1½-inch width by 6 inches
- Paper tape
- Towels, gown to cover patient
- Water: quite warm but not hot to touch
- Bandage scissors

23

Cast Application Instructions

1. Measure stockinette from distal proximal inter-phalangeal (PIP) joints to the acromion process on posterior aspect of arm (Figure 5–1). Apply stockinette.
2. When the elbow is flexed 45° or more, cut a slit on the stockinette at anterior elbow crease horizontally from one epicondyle to the other (Figure 5–2). Overlap the stockinette (Figure 5–3).
3. Apply felt strips:
   a. distal to axilla, circumferentially
   b. over ulnar styloid, circumferentially
   c. across olecranon vertically up on the humerus and down along the ulna
   d. horizontally across olecranon and both epicondyles

   Tape felt in place. Figure 5–4 shows all strips in place.
4. Apply padding. It is easier to begin distally because the circumference of the forearm is smaller distally (Figure 5–5). Wrap circumferentially, overlapping 1½ inches on each wrap. If as you

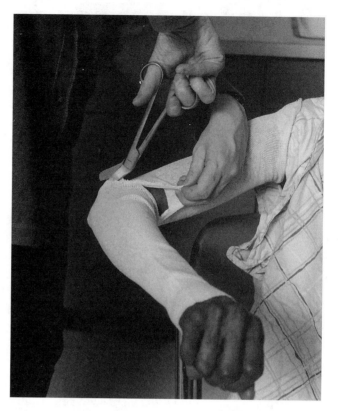

Figure 5–2 Cut a slit at anterior elbow crease.

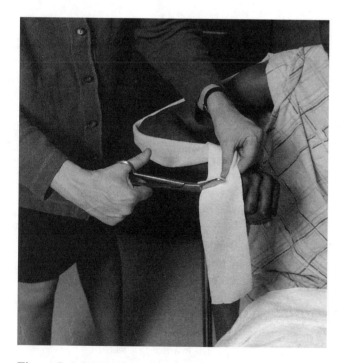

Figure 5–1 Measure stockinette from PIP joint to acromion.

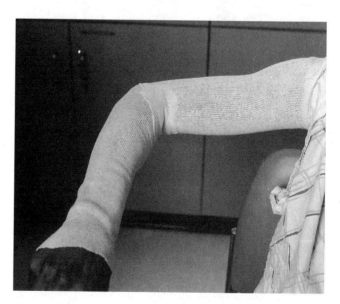

Figure 5–3 Overlap stockinette.

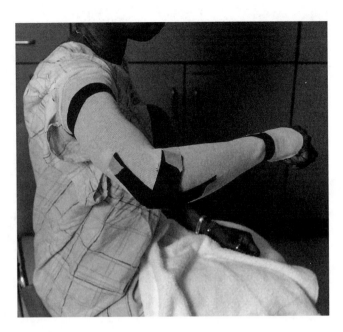

Figure 5–4 Tape felt in place.

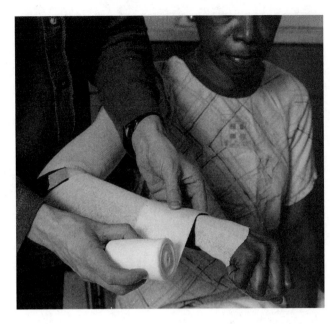

Figure 5–5 Wrap padding circumferentially.

wrap you find the padding bulking on one side, tear the opposite side to make the material conform smoothly to the arm. The padding covers the ulnar styloid distally and extends fully up onto the axilla. Apply five or six layers at both distal and proximal ends and four or five layers covering other areas of the arm.

5. Apply figure-eight wrap at the elbow. Padding should crisscross at anterior elbow crease and overlap approximately 1 inch over the olecranon (Figures 5–6, 5–7, 5–8). Do not bring padding directly over the olecranon as this will cause direct pressure and bulking proximally and distally, and the padding may get stretched too thin at the olecranon.

6. Apply plaster ½ inch below top of padding at the axilla down the humerus to ½ inch proximal to the ulnar styloid. If plaster bulks or narrows, tuck it where it is bulking (Figure 5–9). Rub plaster well into gauze (Figure 5–10).

7. Wash hands and flare proximal and distal edges with a circumferential motion combined with outward pull of the index finger. Do not apply counterpressure with thumb.

8. Pull stockinette over edge of plaster and fix in

place with plaster strips. Completed rigid circular elbow cast is shown in Figure 5–11.

ELBOW DROP-OUT CAST

Rationale for Use

The drop-out cast is so named because either the humeral or forearm portion of the cast is left partially open, allowing the forearm to drop out of the cast as the elbow extends with the pull of gravity and available triceps action. The drop-out cast is used primarily in cases of fixed contracture in a serial casting approach to lengthen the muscles and soft tissues that are contracted.

The drop-out cast is the cast of choice for managing contractures of the elbow flexors that are moderate to severe, or more than 40°. With contractures of 0° to 40° the drop-out cast will slip down over the elbow, potentially causing pressure areas on the skin. With contractures less than 30°, the drop-out cast is likely simply to slip off the arm entirely.

In cases of spasticity as well as contracture, the drop-out cast can be used when the abnormal muscle tone is

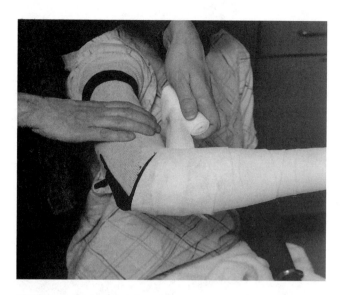

Figure 5–6 Figure-eight wrap at elbow.

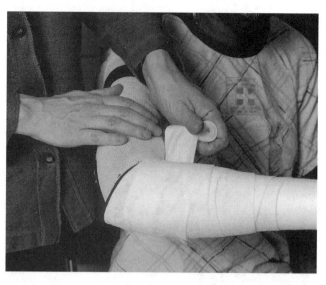

Figure 5–7 Padding crisscrosses at anterior elbow crease.

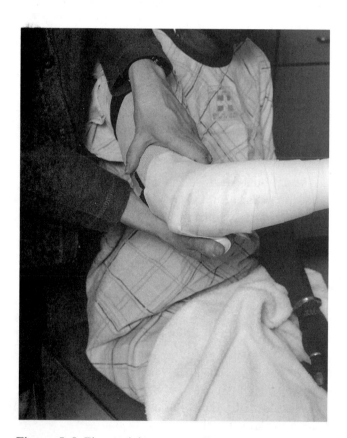

Figure 5–8 Figure-eight wrap at elbow. Padding overlaps about 1 inch over olecranon.

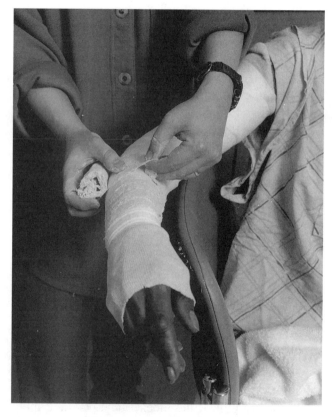

Figure 5–9 Tuck plaster.

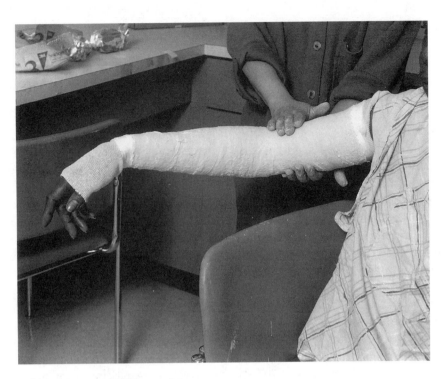

Figure 5–10 Rub plaster well into gauze.

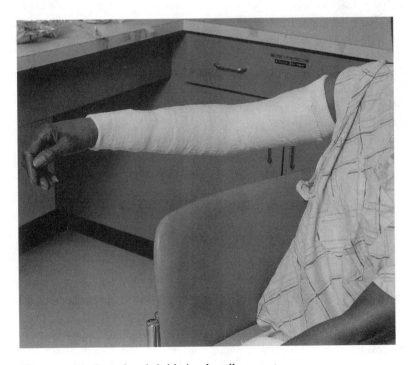

Figure 5–11 Completed rigid circular elbow cast.

mild to moderate and stable. The drop-out cast is not effective when there is significant variation in joint position associated with varying degrees of agitation or relaxation. For example, a patient with an elbow flexion contracture of 50° whose elbow joint position varies from 70° of flexion when he or she is alert and seated properly in a chair to 110° of flexion when agitated, improperly positioned, or in bed supine would not be a good candidate for a drop-out cast. As the patient's resting position changed, he or she would be alternately pulling strongly against the cast when agitated or the cast would be slipping down on the arm when the patient was more relaxed. This would be likely to result in shifting and rubbing of the cast, with significant risk of skin problems developing. In addition, when the patient was pulling strongly against the cast there would be a risk of tendinitis developing. In a case such as this, a more stable cast that does not allow change in joint position, such as the rigid circular elbow cast, would be more appropriate. The drop-out cast would be appropriate for a patient with a 70° elbow flexion contracture and moderate but stable abnormal tone whose resting elbow position varied only 10° to 15° with positional and alertness changes.

The drop-out cast can also be used when there is a problem with skin integrity on the posterior surface of the humerus or elbow, or over the portion of the forearm that is not enclosed. It has also been an effective option for patients who require some daily joint mobilization. This can be the case with patients who have an arthritic condition and also in some cases where heterotopic ossification is present. Heterotopic ossification is generally a contraindication for casting or any form of immobilization. Drop-out casts have been used occasionally even with heterotopic ossification present because slight joint motion can be achieved with this type of cast in place.

For the drop-out cast to be effective, the patient must be positioned to allow gravity to assist in pulling the elbow into extension. This requires that the patient be positioned in an upright position for most of the day. It is also important for the patient to avoid wearing clothing that might restrict the arm from dropping out of the cast. Occasionally it is necessary to combine a wrist hand splint with an elbow cast. When doing so with a drop-out cast, the splint straps should not prevent the arm from dropping out of the cast.

Two types of drop-out casts exist. One encloses the humerus fully from the axilla to the ulnar groove just proximal to the olecranon, then encloses the forearm partially. The portion of the forearm that is covered by the partial gutter or shell depends on the resting position of the forearm. The forearm position cannot be controlled with the drop-out cast. If the forearm rests in neutral, the shell will cover approximately two thirds of the circumference of the forearm, over the radial border and down to the medial border of the ulna on the volar and dorsal forearm surfaces. The opening will be along the ulnar border. If the forearm is resting in supination, the shell will cover the volar surface of the forearm. If the forearm is in pronation, the shell will cover the dorsal forearm surface.

The second type of drop-out cast encloses the full circumference of the forearm and the anterior portion of the humerus. This method has the advantage of the heaviest part of the cast being on the forearm, which is the lever arm gravity is acting on to pull the elbow into extension. A wrist cast can also be combined with the forearm-enclosed drop-out cast to manage elbow and wrist contractures simultaneously. The most significant drawback of the forearm-enclosed drop-out cast is that as the elbow extends and the arm drops out, the humeral portion wedges away from the upper arm and can be cumbersome. Care must be taken that clothing does not prevent the humeral portion from dropping away.

While most casts are left in place for 5 to 7 days, there is more variability with drop-out casts because their effects can be more easily observed while the cast is still in place on the arm. The therapist might observe that 30° of range of motion have been gained in 3 days and opt to change the cast then, rather than waiting for another 2 days. It is also possible that the therapist will observe very little change after 5 days and opt to keep the cast in place longer.

Required Casting Materials

- Stockinette: 3-inch width for most adults; 2-inch width for very small adults and children
- Cast padding: 3-inch width, three to six rolls depending on size of arm
- Plaster: 3-inch width, three rolls, unwrapped, plus seven sets of three strips each. Length of strips is

measured from midhumerus just proximal to the ulnar styloid (for drop-out cast measure on anterior arm; for reverse drop-out cast measure on posterior aspect).
- Felt: four strips:
    one 1½-inch width by circumference of humerus just below axilla
    one 1½-inch width by circumference of forearm over ulnar styloid
    two 1½-inch width by 6 inches
- Paper tape
- Towels, gown to cover patient
- Water: quite warm but not hot to touch
- Bandage scissors

Cast Application Instructions

1. Measure stockinette from the distal metacarpophalangeal (MCP) joints, proximally to the acromion process on the posterior aspect of the arm (Figure 5–12). Apply stockinette. Unroll smoothly on arm. Pulling the stockinette proximally and distally may cause constriction; ensure that the stockinette is as loose as possible without actually having wrinkles.

2. When the elbow is flexed 45° or more, cut a slit on the stockinette at anterior elbow crease, horizontally from one epicondyle to the other (Figure 5–13). Overlap the stockinette (Figure 5–14).
3. Apply felt strips to the following areas:
    a. distal to axilla
    b. over the ulnar styloid
    c. across the olecranon vertically up on the humerus and down along the ulna
    d. horizontally across the olecranon and both epicondyles
    Tape felt in place. Figure 5–15 shows all strips in place.
4. Apply padding (Figure 5–16). It is easier to begin distally because the circumference of the forearm is smaller distally. Wrap circumferentially, overlapping 1½ inches on each wrap. If as you wrap you find the padding bulking on one side, tear the opposite side to make the material conform

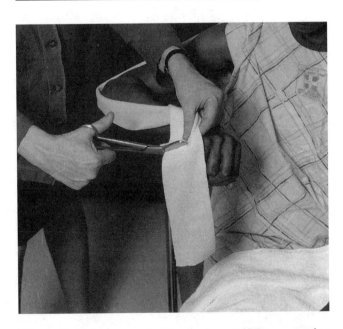

Figure 5–12 Measure stockinette from MCP to acromion process.

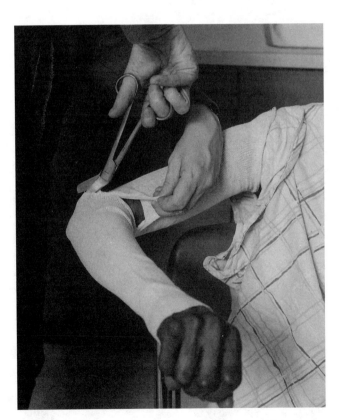

Figure 5–13 Cut a slit on stockinette at anterior elbow crease.

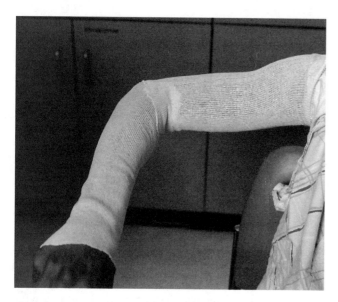

Figure 5–14 Overlap the stockinette.

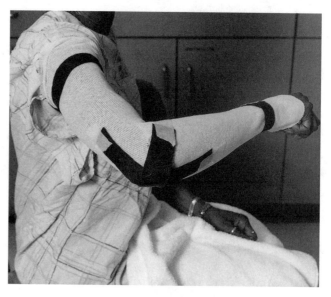

Figure 5–15 Applied felt strips in place.

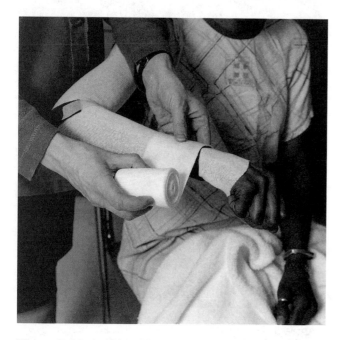

Figure 5–16 Apply padding, starting at distal forearm.

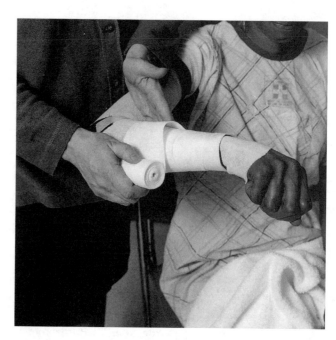

Figure 5–17 Padding is bulking.

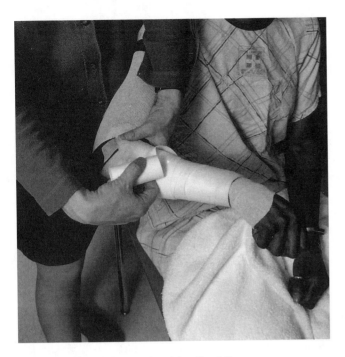

Figure 5–18 Tear opposite side of padding.

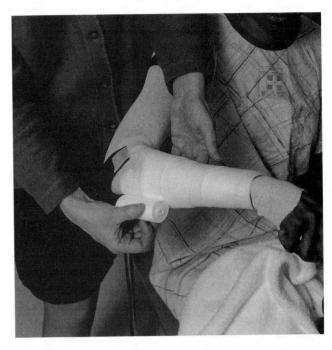

Figure 5–19 Lay padding smoothly on arm.

smoothly to the arm (Figures 5–17, 5–18, 5–19). The padding covers the ulnar styloid distally and extends fully up into the axilla. Apply five or six layers at both distal and proximal ends and four or five layers covering other areas of the arm. Unroll the padding onto the arm; do not pull. When the padding is in place you should be able to get two fingers snugly into cast.

5. Apply figure-eight wrap at the elbow. Padding should crisscross at anterior elbow crease and overlap approximately 1 inch over the olecranon (Figures 5–20, 5–21, 5–22). Do not bring padding directly over the olecranon as this will cause direct pressure, bulking proximally and distally, and the padding may get stretched too thin at the olecranon.

6. If not initially prepared, measure the length for the strips (seven sets of three strips each). Measure length, leaving ½ inch of padding at the proximal and distal ends (Figure 5–23).

7. Apply plaster. Holding onto end of the roll in one hand, immerse the plaster roll in water with the other hand until it bubbles. Lightly squeeze excess water out. Wrap the enclosed portion of the cast from ½ inch below top of padding at axilla down the humerus to the ulnar groove just proximal to the olecranon for the humeral portion enclosed (Figure 5–24). It is important that the full length of the enclosed portion be covered before beginning to apply strips. If plaster bulks or narrows, tuck it where it is bulking (Figure 5–25). Rub plaster well into gauze (Figure 5–26).

8. Apply seven sets of three strips each. Apply the first set on the center of the shell, against the direction the arm tends to pull in. Lay the plaster snugly into the elbow crease (Figure 5–27). The exact position of the shell on the forearm depends on the degree of forearm pronation or supination. If the forearm is fully supinated, the first set is applied from the anterior center humerus across the elbow and down the anterior portion of the forearm. If the forearm is fully pronated, the first set of strips extends down the dorsum of the forearm. If the forearm is in neutral, the plaster strips extend down the radial border of the forearm. Leave ½ inch of padding exposed beyond the plaster out the distal end of the cast. The plaster extends distally to just proximal to the ulnar sty-

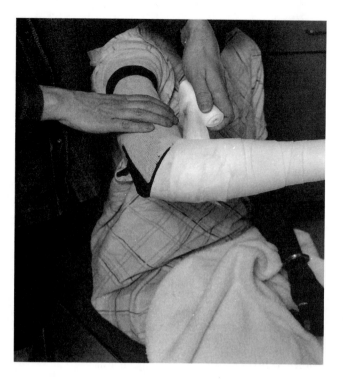

Figure 5–20 Figure-eight wrap at elbow.

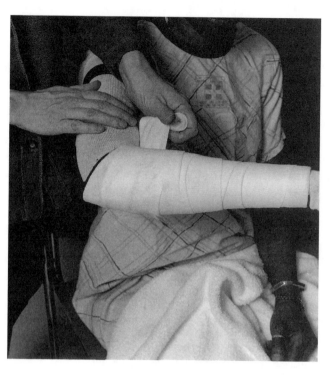

Figure 5–21 Crisscross at anterior elbow crease.

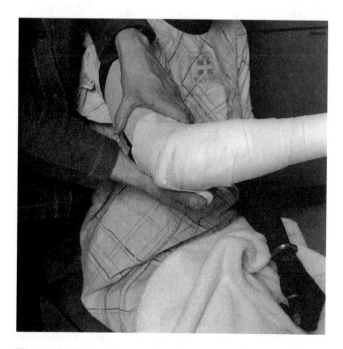

Figure 5–22 Padding overlap 1 inch over olecranon.

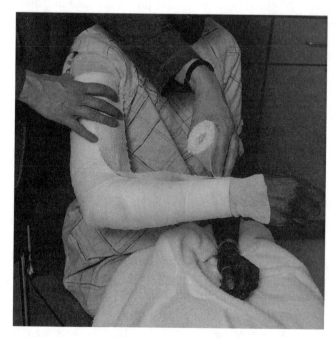

Figure 5–23 Measure length for strips.

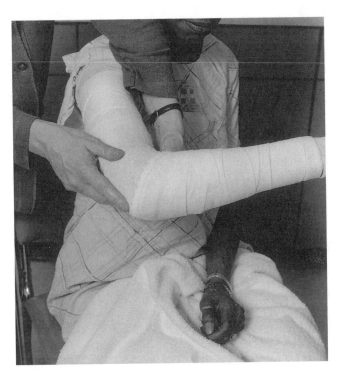

Figure 5–24 Plaster just proximal to olecranon.

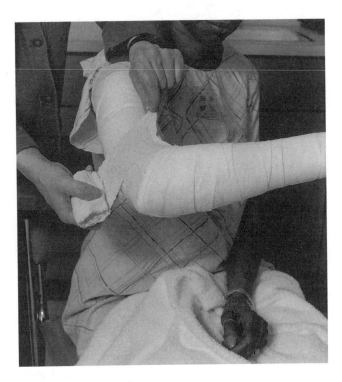

Figure 5–25 Tuck plaster.

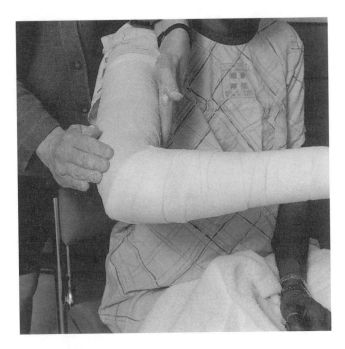

Figure 5–26 Rub plaster into gauze.

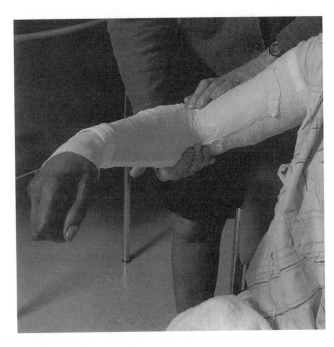

Figure 5–27 First set of three strips.

loid. Apply the second set of strips lateral to the first and angling toward the lateral epicondyle, wrapping around the posterior humerus (Figure 5–28). Apply the third set medially to the first, angling toward the medial epicondyle and wrapping around the posterior humerus (Figure 5–29). The fourth set repeats the first, down the center of the shell (Figure 5–30). The fifth repeats the second, lateral to the first and angling toward the medial epicondyle. The sixth repeats the third, medial to the first and angling toward the medial epicondyle, and the seventh repeats the first, down the center.

9. Wash hands and flare proximal and distal edges with a circumferential motion combined with outward pull of the index finger (Figures 5–31, 5–32). Do not apply counterpressure with thumb. Turn stockinette back over plaster edge and secure with plaster strips.

10. Apply another plaster roll and wrap the enclosed humeral portion again, covering the strips (Figure 5–33). Again attempt to cover the entire length of the enclosed portion with this roll.

11. Finish the shell portion. With bandage scissors, cut distally, going down the center of the padding opposite the shell, gliding the scissors along the stockinette (Figure 5–34). Stop 2 inches from the olecranon (from anterior elbow crease for reverse drop-out cast) (Figure 5–35) and angle toward the lateral epicondyle (Figure 5–36). Then cut toward the medial epicondyle, making a triangle to pull back over the edge of the plaster at the elbow. Cut through the stockinette in the same way (Figure 5–37). Cut away excess padding, leaving ½ inch exposed beyond plaster edge (Figure 5–38). Pull stockinette over edge of the plaster and fix in place with plaster strips.

12. Check that the forearm shell portion of the drop-out cast is not constricting the arm from extending. The drop-out cast should allow the arm to extend as it relaxes to increase range of motion (Figure 5–39). Figure 5–40 shows completed elbow drop-out cast.

RIGID CIRCULAR WRIST CAST

Rationale for Use

In cases of spasticity, flexion is the more dominant function of the hand and may lead to myostatic contrac-

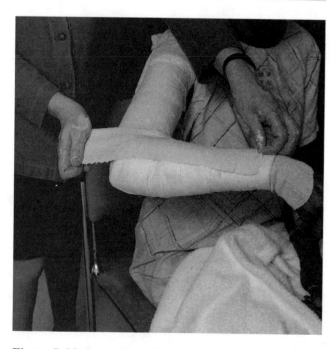

Figure 5–28 Second set of strips.

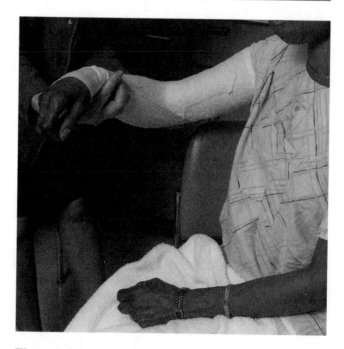

Figure 5–29 Third set of strips.

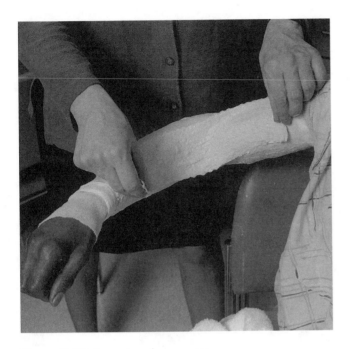

Figure 5–30 Repeat again to center.

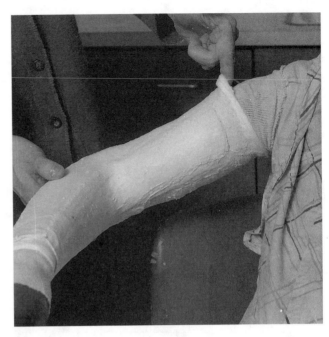

Figure 5–31 Flare proximal edge.

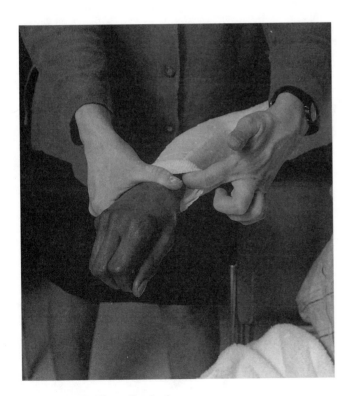

Figure 5–32 Flare distal edge.

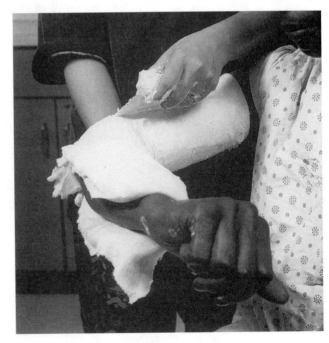

Figure 5–33 Apply another plaster roll.

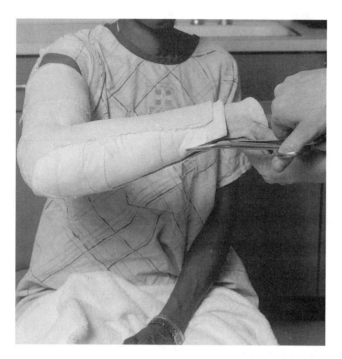

Figure 5–34 Cut padding from distal forearm end.

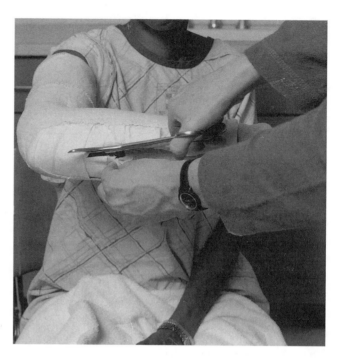

Figure 5–35 Stop 2 inches from olecranon.

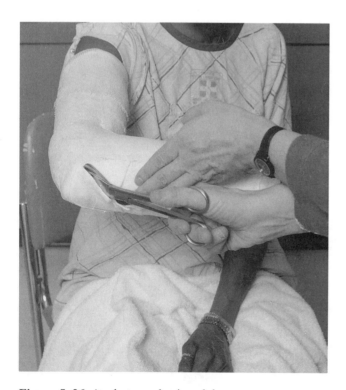

Figure 5–36 Angle toward epicondyle.

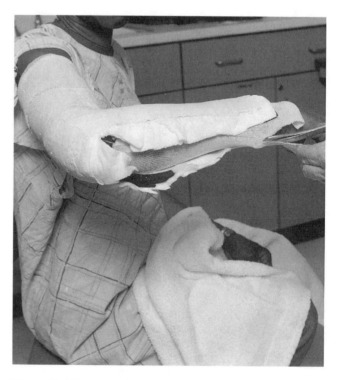

Figure 5–37 Cut stockinette.

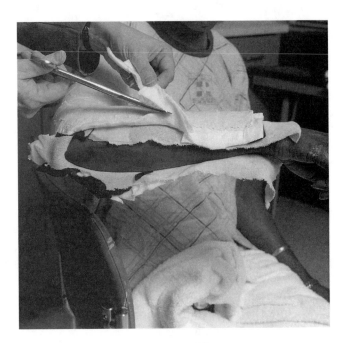

Figure 5–38 Cut away excess padding.

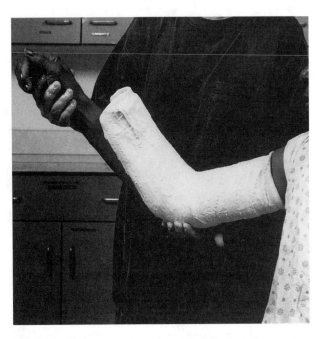

Figure 5–39 Over time, forearm drops out of cast.

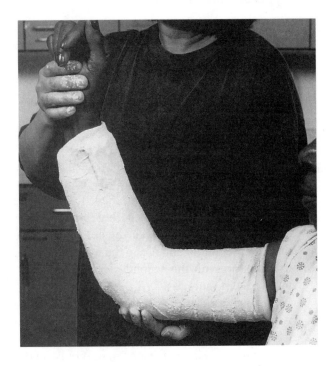

Figure 5–40 Completed elbow drop-out cast.

ture. The muscle imbalance around the wrist creates poor biomechanical alignment, negatively affecting optimal hand function. For example, when the wrist is postured in ulnar deviation, the wrist has a tendency to flex. This in turn places the thumb in poor alignment that will limit the patient's ability to grasp and manipulate.

The balanced alignment of the skeletal structure and muscles that cross the various joints of the wrist and hand create the ability of the hand to move in multiple planes to perform powerful grasp as well as perform intricate movements for precision. The carpal bones serve as a point of stability for the forces of wrist flexors and extensors.

The rigid circular wrist cast is used for those patients who have difficulty with wrist alignment and control. The wrist cast can be positioned in flexion, extension, or deviation depending on the clinical problem. This cast can assist with rebalancing the muscle pull of the wrist flexors or extensors.

By stabilizing the wrist in a functional position, the wrist cast may promote active distal fine motion of the thumb and fingers. The patient can actively use the hand while it is in the cast and this will assist subsequently in achieving muscle rebalancing and carryover into activities of daily living. Gradual improvement may be noted if the patient practices the hand function movement over time.

In the case of a patient with poor opposing muscles that have led to myostatic contracture or severe wrist flexion, a series of wrist casts to increase gradually the passive range of motion may be required. The hand may be tightly fisted, with the wrist, palm, and fingers dominated by flexion spasticity, and it may have limited ability to expand. The tightly fisted hand pulls the wrist and hand joints out of alignment. This limits the ability to control functional and voluntary movements. Serial wrist casting can improve passive or active range of motion in the hand and possibly improve the potential for functional use. In addition, the improvement in range of motion can allow the therapist to fabricate a follow-up orthosis that is more conducive to good skin and hygiene care. If there is active finger motion and no functional wrist movement, this cast may be especially helpful in assisting the physician and therapist in determining surgical options.

During cast application the therapist should mold and contour the cast to support the wrist, lengthen the shortened muscle, and facilitate fine prehension. Follow the contours of the hand and forearm, placing minimal stretch on the joint and muscles, working in submaximal range. The cast application is similar to that of a hand splint, generally two thirds the length of the forearm to provide adequate leverage. There should be no blanched areas and no pressure areas over bony prominences such as the ulnar styloid. The palmar arch is formed and shaped into a gutter while shaping the palm of the wrist cast. The proximal palmar crease that begins on the radial side of the palm should be the guideline, as in hand splinting, to allow for MCP flexion. The crease slant should be formed diagonally across the palm of the hand, not straight. To create maximal mobility within optimal stability in the cast, it is important to allow the fourth and fifth fingers to work as a stabilizing unit while the thumb, index, and middle fingers provide the precision pinch.

The basic wrist cast can be modified to manage problems seen in the wrist in combination with involvement of the hand, fingers, or thumb. The wrist cast is generally left in place for 5 to 7 days unless con-

traindicated, such as for a patient with juvenile rheumatoid arthritis. The cast should be reevaluated according to need and purpose.

Required Casting Materials

- Stockinette: 3-inch width for most adults; 2-inch width for children
- Cast padding: 3-inch width, three to five rolls depending on size of arm; 2-inch width for children
- Plaster: 3-inch width, four to five rolls depending on size of arm; 2-inch width for children
- Felt: four strips:
  one 1½-inch width by circumference of forearm below elbow joint
  one 1½-inch width by circumference of forearm over ulnar styloid
  one 1½-inch width by 5 inches
  one 1-inch width by 3 inches
- Paper tape
- Towels, gown to cover patient
- Water: quite warm but not hot to touch
- Bandage scissors

Cast Application Instructions

1. Measure stockinette from the olecranon to the PIP joints (Figure 5–41) and approximate placement of the thumb. Cut a straight slit (¼ inch) for the thumb (Figure 5–42) and apply stockinette (Figure 5–43).
2. Apply felt strips to the following:
   a. distal to the olecranon
   b. over the ulnar styloid (Figure 5–44)
   Use the two small strips for the thumb piece. Fold the larger strip in half and cut a heart shape one third from the top (Figure 5–45). Place the smaller felt strip through the hole of the larger strip (Figure 5–46). Place the felt strip piece through the thumb and along the radial border of the forearm with the second piece covering the web space (Figure 5–47). Tape felt in place.
3. Apply padding. It is easier to begin distally at the wrist, because the circumference of the forearm is smaller (Figure 5–48). Wrap circumferentially, overlapping 1½ inch on each wrap. If bulking occurs at the narrow end while wrapping circumferentially (Figure 5–49), hold padding close to

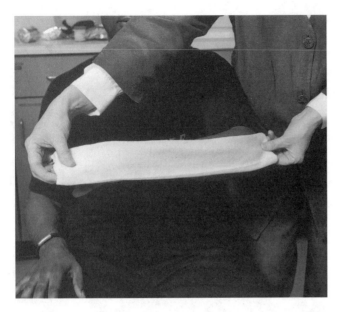

Figure 5–41 Measure stockinette from the olecranon to the PIP joints.

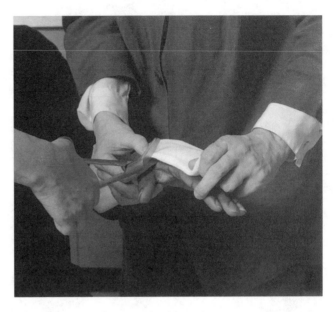

Figure 5–42 Cut a straight slit for the thumb.

Figure 5–43 Apply stockinette.

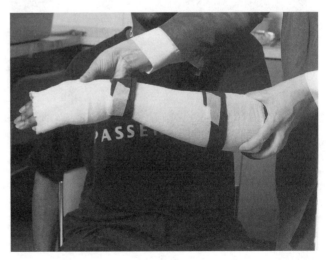

Figure 5–44 Apply felt strip distal to olecranon and over ulnar styloid.

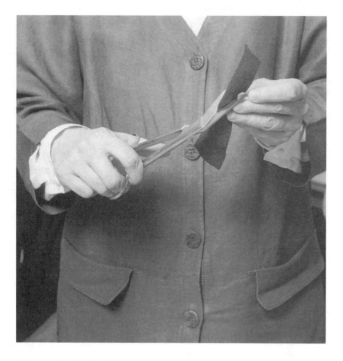

Figure 5–45 Fold larger strip in half and cut a heart shape.

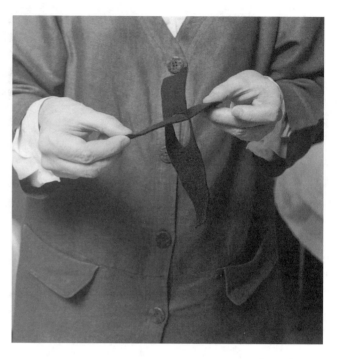

Figure 5–46 Place smaller felt strip through hole of larger felt strip.

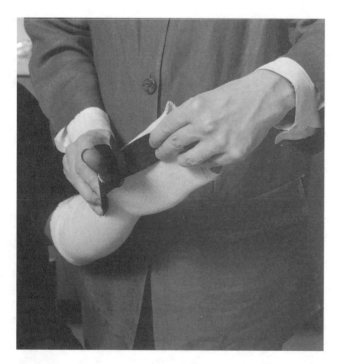

Figure 5–47 Place felt strip piece through thumb, covering web space.

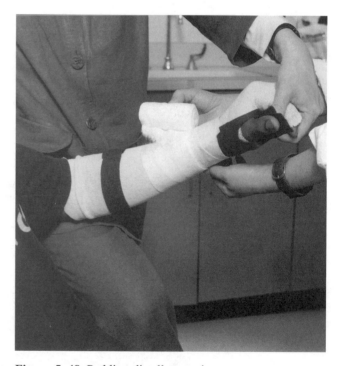

Figure 5–48 Padding distally at wrist.

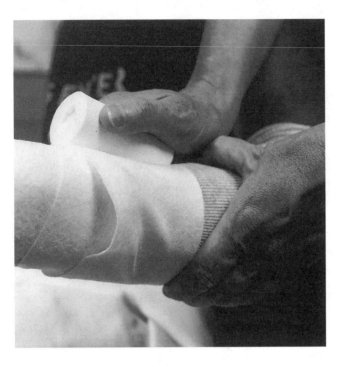

Figure 5–49 Bulking of padding at narrow end.

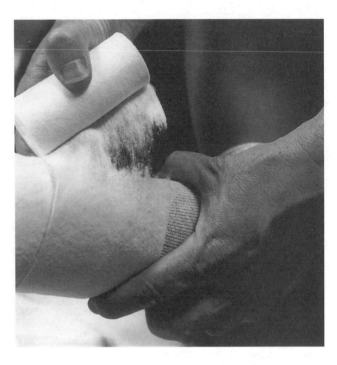

Figure 5–50 Tear opposite bulking end to contour.

the forearm and tear at the opposite end to contour to the shape of the forearm (Figure 5–50). The padding covers the ulnar styloid distally and extends fully up three fourths the length of the forearm (Figure 5–51) with four to five layers at the proximal and distal ends.

4. When wrapping distally at the hand, apply figure-eight wrap around the thumb. This technique is to ensure that there is padding over both the thumb and the second metacarpal surfaces of the web space to prevent skin breakdown. Unroll padding on the volar surface of the hand ulnar to radial so that the material is pulled up through the web space into wrist extension, and radial to ulnar on the dorsal surface. Apply the figure-eight technique by using either of the following methods:

Method A

Aim middle roll of padding to middle of web space (Figure 5–52). Tear padding vertically at the third or fourth metacarpal on the volar surface (Figure 5–53). Pull padding up through web space (Figure 5–54). Tear vertically again on dorsal surface at the second metacarpal (Figure 5–

55). Wrap ends back around the MCP joint of thumb (Figure 5–56). Finally, wrap back around carpometacarpal (CMC) joint ulnar to radial on volar surface, radial to ulnar on dorsal surface (Figure 5–57). Repeat.

Method B

Aim middle roll of padding to middle of web space (Figure 5–58). As padding is pulled up through web space, tear horizontally at the third metacarpal on dorsal surface (Figure 5–59). Wrap bottom half of padding around metacarpal of thumb and lay the other half of padding across the MCP joints (Figure 5–60). Finally, wrap back around CMC joint, ulnar to radial on volar surface, radial to ulnar on dorsal surface. Repeat.

5. Apply plaster ½ inch below the top of padding at proximal end of forearm and unroll, leaving ½ inch of padding at the distal end. If bulking occurs with the plaster (Figure 5–61), take the extra plaster and tuck (Figures 5–62a and 5–62b). Apply figure-eight wrap around the thumb, ulnar to radial on the volar surface of the hand. Squeeze plaster together at the web space so that it lays in

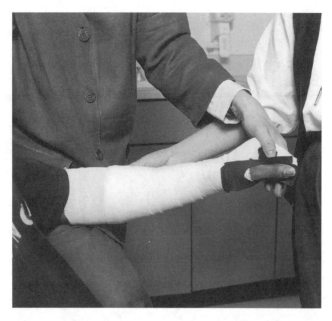

Figure 5–51 Padding covering three-fourths length of forearm.

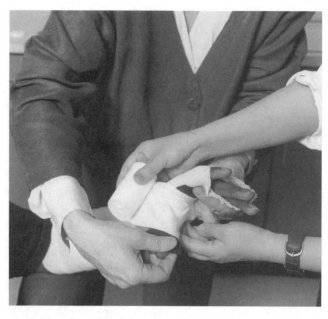

Figure 5–52 Aim middle roll of padding to middle of web space.

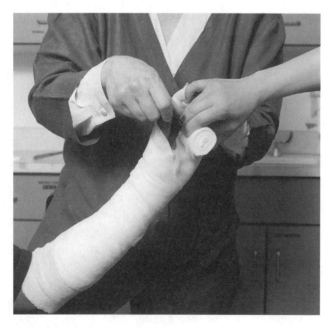

Figure 5–53 Tear padding vertically at third or fourth metacarpal on volar surface.

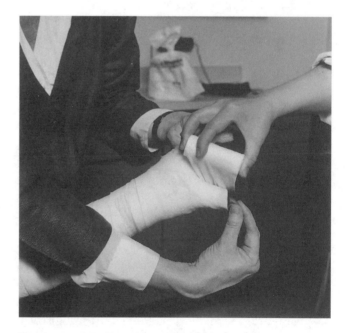

Figure 5–54 Pull up through web space.

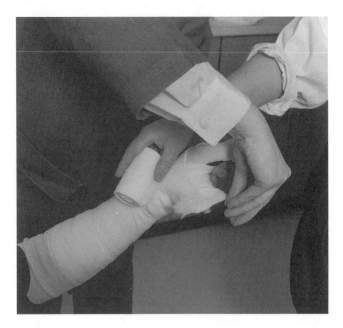

Figure 5–55 Tear vertically on dorsal surface at second metacarpal.

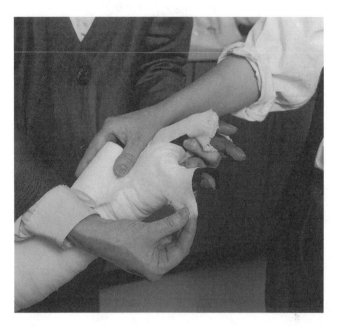

Figure 5–56 Wrap ends back around MCP joint of thumb.

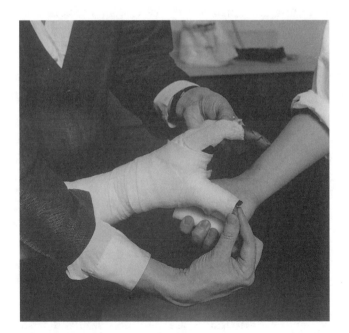

Figure 5–57 Wrap back around CMC and MCP joints of thumb.

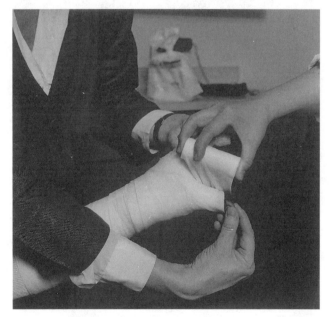

Figure 5–58 Aim middle roll of padding to web space.

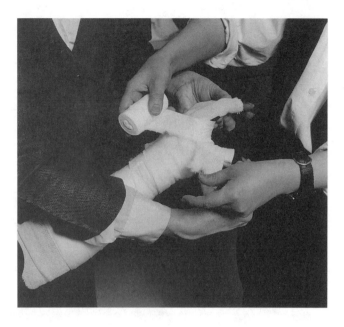

Figure 5–59 Tear horizontally at third metacarpal on dorsal surface.

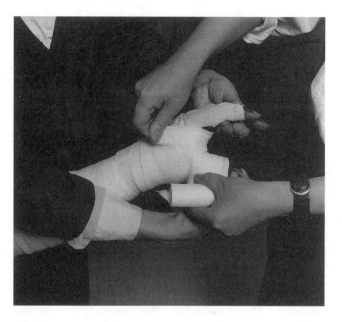

Figure 5–60 Padding around metacarpal of thumb and other half across MCPs of fingers.

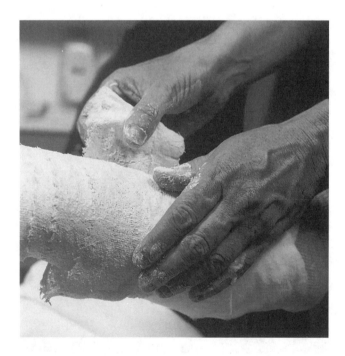

Figure 5–61 Bulking of plaster.

the trough formed by the padding (Figure 5–63). Angle the plaster and tuck to continue unrolling radial to ulnar on the dorsum of the hand and back around the volar surface of the wrist (Figure 5–64). Come through the web space with the plaster only three times, or it may become too bulky.

6. After applying one to two rolls of plaster, form a palmar arch. Apply pressure on the volar surface with index, middle, and ring fingers and counterpressure at the wrist, using the volar surface of your hand (Figure 5–65). Do not apply counterpressure with the thumb or fingertips. Apply additional layers of plaster to reinforce cast if needed.

7. Turn stockinette back over plaster edge and secure with plaster strips. Finish the thumb area by folding the felt and secure with the stockinette for the web space.

8. Wash hands and flare proximal and distal edge with a circumferential motion combined with outward pull of the index finger. Flare the MCP area of the thumb. Do not apply counterpressure

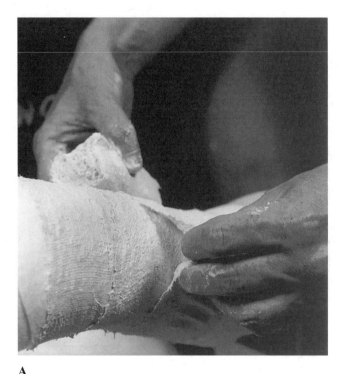

A

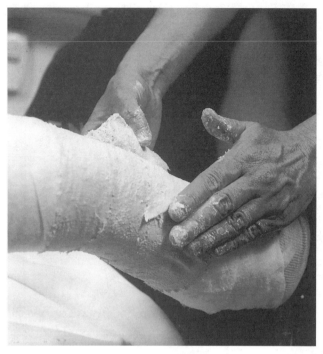

B

Figure 5–62 Tuck of plaster.

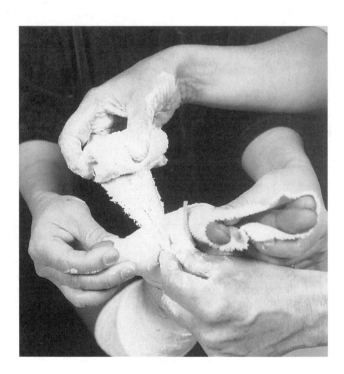

Figure 5–63 Squeeze plaster together at web space.

with the thumb. To check for tightness at the distal end, shift the fingers in the cast in a radial direction and place the index finger into the ulnar edge of the cast. Shift fingers in an ulnar direction and place index finger in the radial edge of the cast. Figure 5–66 shows the completed wrist cast.

LONG ARM CAST

Rationale for Use

A very common pattern in patients with spasticity in the upper extremity is abduction and internal rotation of the humerus, flexion of the elbow, pronation of the forearm, and flexion of the wrist and fingers. The ability of the patient to rotate the forearm to orient the hand for fine motor activities is diminished if the patient cannot actively supinate. In addition, the malalignment of the joints and muscle tightness affect the quality of the hand movement and prevent activities that require fine prehension.

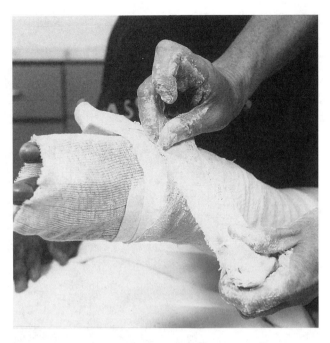

Figure 5–64 Angle plaster and tuck to continue.

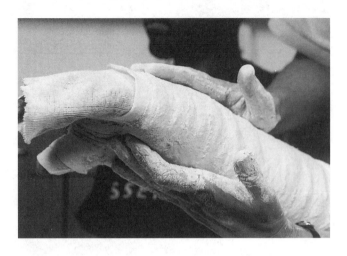

Figure 5–65 Form palmar arch.

The long arm cast is effective in controlling the position of the forearm by increasing the degrees of supination or pronation. It includes the elbow, forearm, and wrist. The cast consists of the proximal portion of the humerus and the distal wrist portion.

The long arm cast is designed to apply equalized pressure in the forearm motion, which takes place equally in both the proximal and distal radioulnar joint. The wrist motion has been included to assist with controlling the amount of the forearm rotation as well as elbow extension. The wrist is generally positioned in neutral, since forearm rotation influences carpal alignment.

For patients with mild to moderate spasticity, the long arm cast may reduce the dominance of the spasticity and assist in rebalancing the antagonist motor group. The position of the long arm cast will place the muscles of the elbow and forearm at a biomechanical advantage for balanced activity. While wearing the long arm cast, patients may improve scapular control as they increase the strength of their shoulder girdle musculature.

The long arm cast is contraindicated for patients with moderate to severe flexor spasticity. The cast crosses multiple joints, which may cause overstretch-ing, skin breakdown, and microtearing of the soft tissue. Initially the tightly flexed arm should be casted proximally at the elbow using the rigid circular elbow cast, working on one joint at a time.

In a case of muscle paralysis, such as that in a patient with spinal cord injury, an unopposed biceps may develop into a supination contracture as the forearm is pulled into supination during elbow flexion. An early rehabilitation program is essential to prevent this deformity. When there is a lack of hand grasp and release, use of a hand orthosis is required to facilitate a functional tenodesis. In this scenario, the position of full pronation is critical. The long arm cast can position the forearm into pronation with the elbow extended to relax the biceps and its pull into supination gradually.

During cast application, the therapist first gently holds the involved hand, positions the elbow and forearm, and feels for the reduction of the tension. Generally, this is at the submaximal range. The therapist unrolls the padding and plaster, applying equalized pressure throughout the length of the arm. Using equal constant force throughout the arm, the overall result will be gradual decrease in the abnormal tone and improvement in the range. Muscle reeducation will be important for facilitating isolated control, since muscle weakness is common.

Required Casting Materials

- Stockinette: 3-inch width for most adults; 2-inch width for very small adults and children
- Cast padding: 3-inch width, three to six rolls depending on size of the arm
- Plaster: 3-inch width, four to six rolls depending on size of arm
- Felt: six strips
  one 1½-inch width by circumference of humerus just below axilla
  one 1½-inch width by circumference of forearm over ulnar styloid
  two 1½-inch width by 6 inches
  one 1½-inch width by 5 inches
  one 1-inch width by 3 inches
- Paper tape
- Towels, gown to cover patient
- Water: quite warm but not hot to touch
- Bandage scissors

Cast Application Instructions

1. Measure stockinette from fingertips to acromion process on posterior aspect of arm (Figure 5–67), and approximate placement of the thumb. Cut a straight slit for the thumb (Figure 5–68). Apply stockinette (Figure 5–69). When the elbow is flexed 45° or more, cut a slit on the stockinette at anterior elbow crease, horizontally from one epicondyle to the other (Figure 5–70). Overlap the stockinette.
2. Apply felt strips to the following areas, similar to elbow and wrist cast (Figure 5–71):
   a. distal to the axilla
   b. over the ulnar styloid
   c. across olecranon vertically up on the humerus and down along the ulna
   d. horizontally across olecranon and both epicondyles

   Two strips are required for thumb piece. Fold the larger strip in half and cut a heart shape one third from the top. Place the smaller felt strip through the hole of the larger strip. Place the felt strip piece through the thumb and along the radial bor-

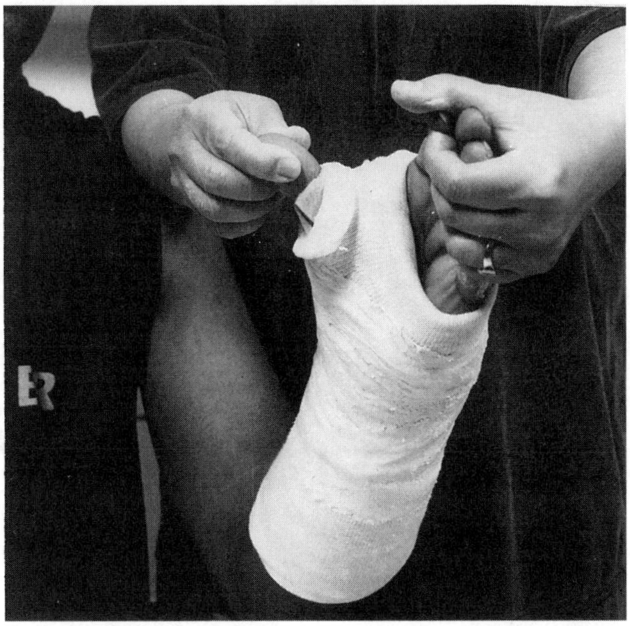

Figure 5–66 Completed rigid circular wrist cast.

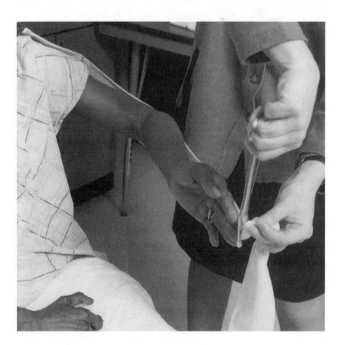

Figure 5–67 Measure stockinette from fingertips to acromion.

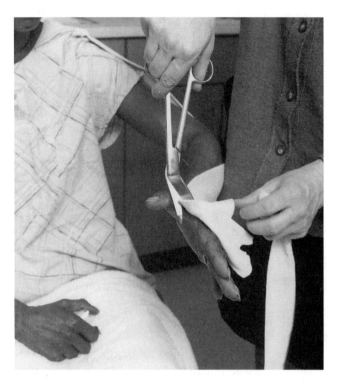

Figure 5–68 Cut slit for thumb.

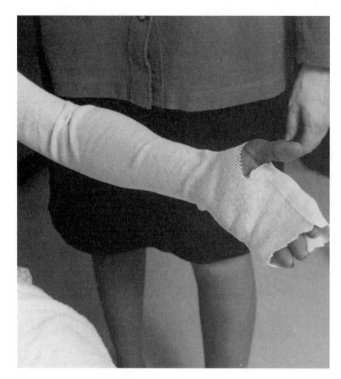

Figure 5–69 Applied stockinette.

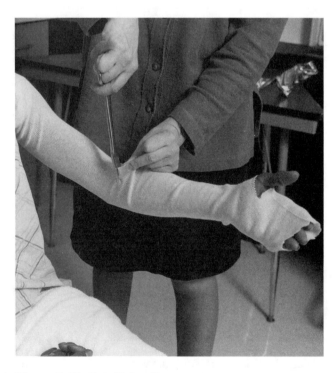

Figure 5–70 Cut slit for elbow crease.

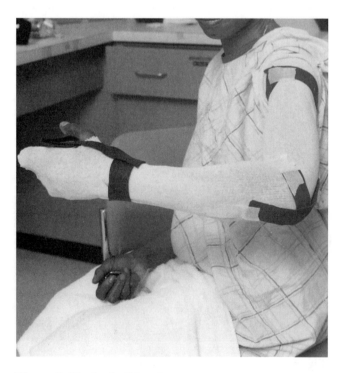

Figure 5–71 Apply felt strips.

der of the forearm, with the second piece covering the web space. Tape felt in place.

3. Apply padding as described for the elbow and wrist cast. It is easier to begin distally because the circumference of the forearm is smaller distally. Wrap circumferentially, overlapping 1½ inches on each wrap. If as you wrap you find the padding bulking on one side, tear the opposite side to make the material conform smoothly to the arm. The padding covers the hand distally and extends fully up onto the axilla. Apply five to six layers at both distal and proximal ends and four to five layers covering other areas of the arm.

4. Apply figure-eight wrap around the elbow as described in the rigid circular elbow cast procedure. Padding should crisscross at the anterior elbow crease and overlap approximately 1 inch over the olecranon. Do not bring padding directly over the olecranon as this will cause pressure at that point, bulking proximally and distally, and the padding may get stretched too thin at the olecranon.

5. Apply figure-eight wrap around the thumb. As described in the wrist cast procedure use either of the following methods:

Method A

Aim middle roll of padding to middle of web space. Tear padding vertically at the third or fourth metacarpal on the volar surface. Pull padding up through web space. Tear vertically again on dorsal surface at the second metacarpal. Wrap ends back around the MCP joint of the thumb. Wrap back around CMC joint ulnar to radial on the volar surface, radial to ulnar on the dorsal surface. Repeat.

Method B

Aim middle roll of padding to middle of web space. As padding is pulled up through the web space, tear vertically at the third metacarpal on the dorsal surface. Wrap proximal half of padding around the metacarpal of the thumb and lay the other half of padding across the MCP joints. Wrap back around the CMC joints ulnar to radial on the volar surface, radial to ulnar on the dorsal surface. Repeat.

6. Apply plaster ½ inch below the top of padding at the axilla, leaving ½ inch at the distal end. Follow procedure for elbow (see figure-eight plaster technique; Figure 5–72) and wrist cast. If plaster bulks or narrows, tuck it where it is bulking (Figure 5–73). Apply figure-eight around the thumb, ulnar to radial on the volar surface of the hand. Squeeze plaster together at the web space so that it lays in the trough formed by the padding as described in the rigid circular wrist cast procedure. Angle the plaster and tuck to continue unrolling radial to ulnar on the dorsum of the hand and back around the volar surface of the wrist. Come through the web space with the plaster only three times or it may become too bulky.

7. After applying two to three rolls of plaster down the arm and hand, form a palmar arch (Figure 5–74). Apply pressure on the volar surface with index, middle, and ring fingers and counterpressure at the wrist using the volar surface of your hand. Do not apply counterpressure with thumb or fingertips. Apply additional layers of plaster to reinforce cast if needed.

8. Wash hands and flare proximal and distal edges with a circumferential motion combined with outward pull of the index finger. Flare the MCP area of the thumb (Figure 5–75). Do not apply counterpressure with thumb. To check for tightness at the distal end, shift the fingers in the cast in a radial direction and place your index finger into the ulnar edge of the cast (Figure 5–76). Shift fingers in an ulnar direction and place your index finger in the radial edge of cast.

9. Turn stockinette back over plaster edge and secure with plaster strips. Finish the thumb area by folding the felt and secure with the stockinette for the web space. Figure 5–77 shows the completed long arm cast.

FINGER SHELL CAST

Rationale for Use

Flexion deformities of the wrist and fingers are the most common upper extremity disabilities in patients with severe to moderate spasticity. The tightness of the wrist and fingers not only prevents extension of the wrist, but renders the hand ineffective in grasp and release because of the overstretched extensor mechanism and shortened flexor muscles of the wrist and fingers.

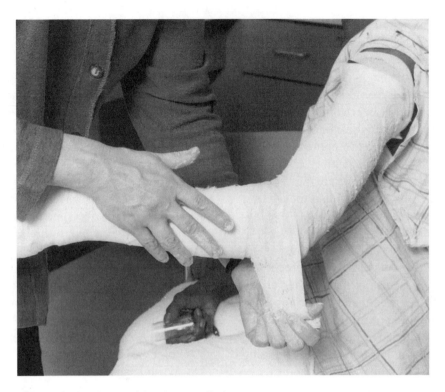

Figure 5–72 Figure-eight plaster technique.

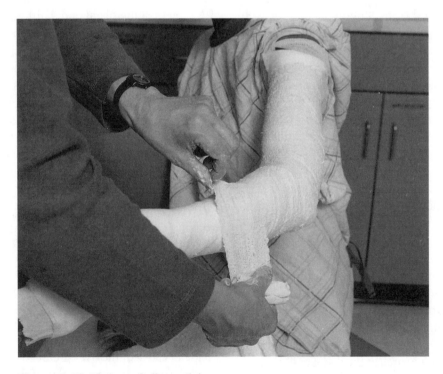

Figure 5–73 If plaster bulks, tuck it.

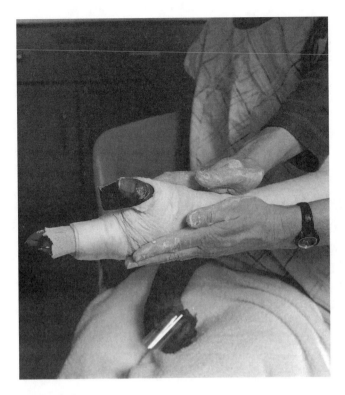

Figure 5–74 Form palmar arch.

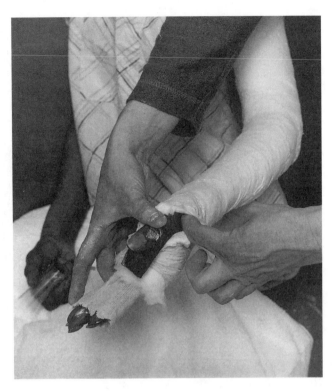

Figure 5–75 Flare MCP area.

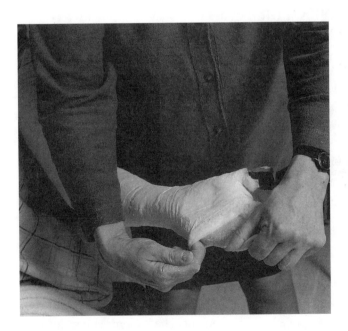

Figure 5–76 Shift fingers in cast and check.

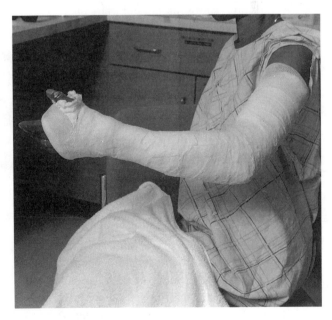

Figure 5–77 Completed long arm cast.

A wrist flexion deformity is often due to a combination of wrist and finger flexor spasticity. Myostatic contractures of the long finger flexors and tightness of the intrinsic muscles can interfere with hand function as well as hygienic care of the hand.

As mentioned, in cases of moderate to severe spasticity the rigid circular wrist cast can be applied in a series as a method for gradually improving the range of wrist extension. However, when there is a more severe wrist-finger flexion deformity, a finger shell can be attached to the rigid wrist cast. The finger shell is removable so that the fingers can be monitored for complications and to allow for a gradual buildup of wearing tolerance. Care must be taken to provide a slow, gradual elongation of the intrinsic muscles of the hand and fingers to prevent microtearing. As the fingers relax, the shell can be progressively positioned into extension. As the range improves in the wrist and fingers, the patient may require a follow-up hand splint.

Required Casting Materials

- Stockinette: 2-inch width for most adults
- Cast padding: 1-inch width, 2- to 3-foot strip
- Plaster: 1-inch width, 2- to 3-foot strip
- Velcro: 1-inch by 12-inch loop, 1- by 4-inch hook
- Towels
- Water: quite warm but not hot to touch
- Bandage scissors

Cast Application Instructions

1. Fabricate wrist cast.
2. Measure stockinette from midmetacarpals to 1 inch past fingertips. Apply stockinette. Holder holds from volar fingertips only under stockinette (Figure 5–78).
3. Wrap 1-inch padding circumferentially proximally at the MCP joints (Figure 5–79). Wrap in figure-eight fashion at the PIP joints to avoid direct pressure over these joints (Figure 5–80). Continue to wrap to the distal fingertips. Use four to five layers of padding.
4. Wrap 1-inch plaster from the distal MCP joints (Figure 5–81) to the fingertips, leaving ½ inch of padding exposed, proximal and distal to the fin-

gers. Fashion a figure-eight around the flexed finger joints.
5. Conform the shell using one finger under the PIP joints on the volar surface. Apply counterpressure on the dorsum with one finger over the proximal and one over the middle phalanges (Figure 5–82).
6. Turn stockinette back over plaster edge and secure with plaster strips (Figure 5–83).
7. Check tightness of the finger shell. Shift fingers in a radial direction and place a finger inside the ulnar border. Shift fingers in an ulnar direction and place a finger inside the radial border. The finger shell should be easily removable.
8. Lay one end of the Velcro loop side facing up on the proximal dorsal surface of the finger shell and the remaining Velcro hanging over the distal fingers (Figure 5–84).
9. Plaster Velcro into shell (Figure 5–85).
10. Pull Velcro loop back on top of the shell with the loop side facing down (Figure 5–86), and plaster over again.
11. Position a Velcro hook (Figure 5–87). Plaster it into the wrist cast (Figure 5–88). The distal end of the Velcro loop can then be attached to a Velcro hook on the wrist cast (Figure 5–89).

PLATFORM CAST

Rationale for Use

For someone to be able to reach out and hold a hand position, a balanced set of coaxial forces is exerted during both lengthening and shortening of the wrist and finger muscles. Passively, as the wrist is extended, the increased tension in the long finger flexors results in increased finger flexion. Conversely, as the wrist is flexed a tenodesis effect results in extension of all the finger joints. The carpal joint and MCP joints must be stable so that the fingers can effectively flex and extend, allowing the hand to accommodate for various sizes of objects.

Hand function in patients with mild spasticity presents therapists with an interesting challenge. They must build in appropriate stabilizing and mobilizing components of movement. The patient with mild spasticity is generally able to isolate wrist and finger

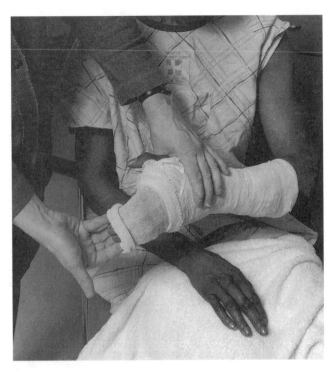

Figure 5–78 Apply stockinette. Holder holds from volar fingertips.

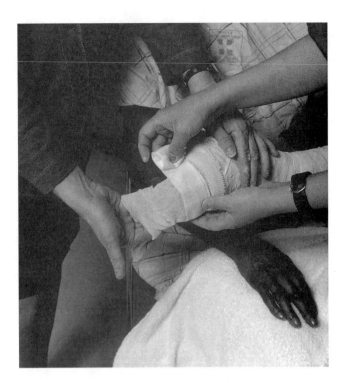

Figure 5–79 Wrap 1-inch padding.

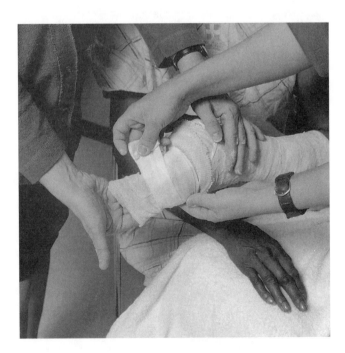

Figure 5–80 Figure-eight wrap at PIP joints.

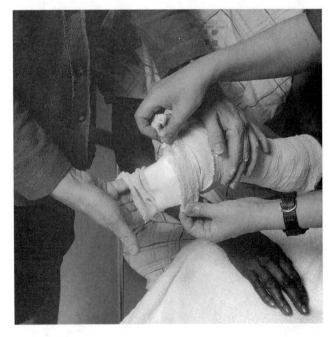

Figure 5–81 Wrap 1-inch plaster from distal MCP joints to fingertips.

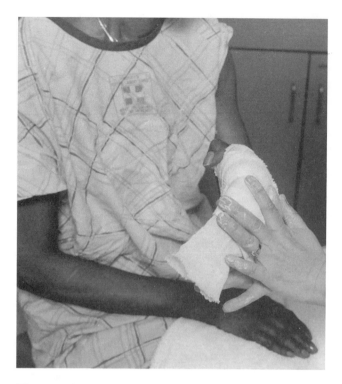

Figure 5–82 Conform the shell.

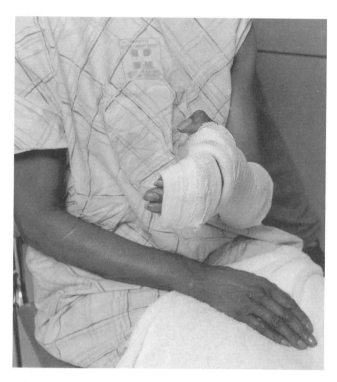

Figure 5–83 Stockinette turned back with finished edge.

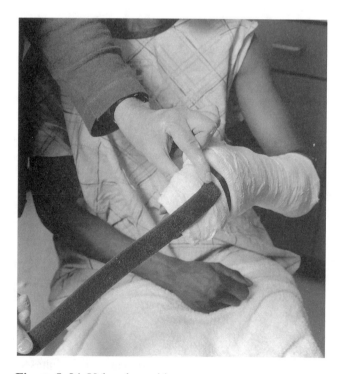

Figure 5–84 Velcro loop side up.

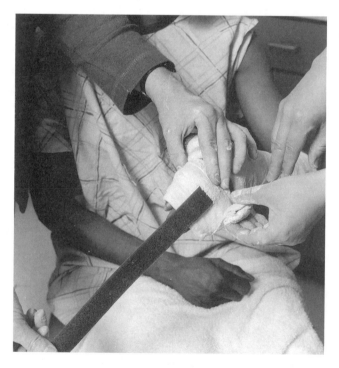

Figure 5–85 Plaster Velcro into shell.

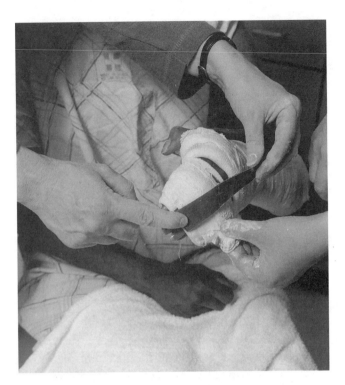

Figure 5–86 Pull loop back and plaster again.

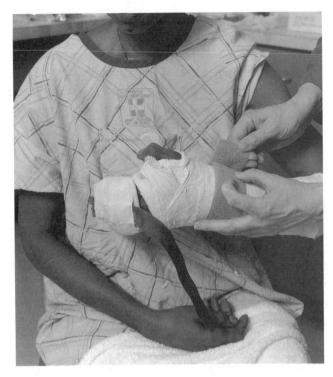

Figure 5–87 Position Velcro hook into wrist cast.

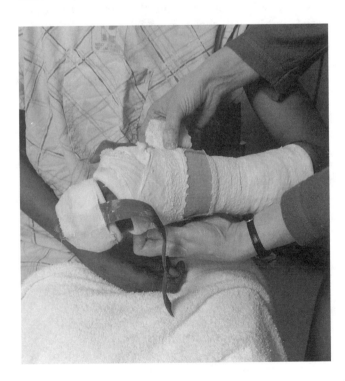

Figure 5–88 Plaster Velcro into wrist cast.

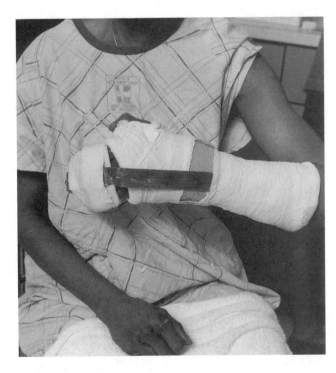

Figure 5–89 Completed finger shell attached to wrist cast.

movements, but has not entirely overcome the influence of flexion. The wrist typically is positioned slightly below neutral or 20° of flexion when the thumb and fingers are used for fine prehension. As the patient attempts to extend the fingers or to incorporate the thumb in prehensile and manipulative actions, the wrist will extend beyond neutral and a corresponding shortening of the long finger flexors and thumb flexors will become evident.

A series of platform casts can be used to improve simultaneous wrist and finger motion gradually in a patient with mild spasticity. This cast is designed to simulate the position the hand would assume in a weight-bearing situation. The platform cast acts to expand the width of the palm while lengthening the long finger flexors. The extent of wrist extension will vary depending on the tightness of the finger flexors.

Caution must be taken when positioning the wrist and fingers in the cast. The finger platform section should have the sides and the volar surface contoured. The latter is necessary to avoid a boutonniere effect at the PIP and distal interphalangeal (DIP) joints. In addition, extreme range into wrist extension should be avoided. The therapist must observe the precaution of casting at submaximal range.

The platform cast can be used to position the wrist and fingers gradually into extension to facilitate a more active and controlled grasp and release mechanism. The platform cast can then be applied with or without the thumb enclosed.

The materials and cast application procedures for the platform cast are similar to those for the rigid circular wrist cast.

Required Casting Materials

- Stockinette: 3-inch width for most adults
- Cast padding: 3-inch width, three to five rolls depending on size of arm
- Plaster: 3-inch width, four to five rolls depending on size of arm, plus three sets of three strips each. The length of strips is measured from midforearm distally to fingertips.
- Felt: four strips
  one 1½-inch width by circumference of forearm below elbow joint
  one 1½-inch width by circumference of forearm over ulnar styloid
  one 1½-inch width by 5 inches
  one 1-inch width by 3 inches
- Paper tape
- Towel, gown to cover patient
- Water: quite warm but not hot to touch
- Bandage scissors

Cast Application Instructions

1. Measure stockinette from olecranon to 1 inch past the fingertips, and approximate placement of the thumb (Figure 5–90). After cutting a straight slit for the thumb, apply stockinette (Figure 5–91).

2. Apply felt strips as previously described for the rigid circular wrist cast:
   a. distal to the olecranon
   b. over the ulnar styloid
   Use the two small strips for the thumb piece. Fold larger strip in half and cut a heart shape one third from the tip. Place the smaller felt strip through the hole of the larger strip. Place the felt strip piece through the thumb and along the radial border of the forearm, with the second piece covering the web space. Tape felt in place (Figure 5–92).

3. Apply padding. Use the same procedure as for rigid wrist cast; however, continue to apply the padding distally to cover 1 inch past the fingertips. Holder can hold only from tips of fingers under the stockinette. It is easier to begin distally at the wrist, because the circumference of the forearm is smaller. Wrap circumferentially, overlapping 1½ inch on each wrap. If bulking occurs at the narrow end while wrapping circumferentially, hold padding close to forearm and tear at the opposite end to contour to the shape of the forearm. The padding extends fully three fourths the length of the forearm and past the distal DIP joint. Apply four to five layers at both proximal and distal ends. The finished product should look like that in Figure 5–93.

4. When wrapping distally at the hand, figure-eight wrap around the thumb. As described in the wrist cast procedure, use either of the following methods:

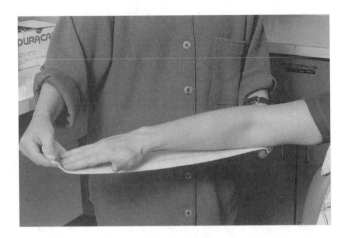

Figure 5–90 Measure stockinette from olecranon to 1 inch past fingertips.

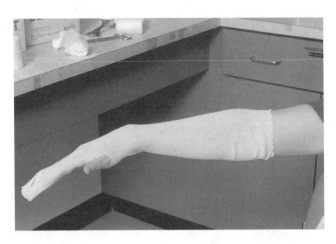

Figure 5–91 After cutting slit for thumb, apply stockinette.

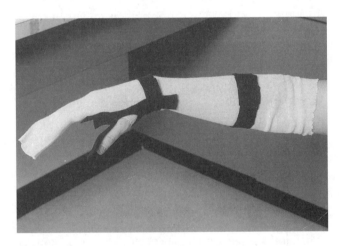

Figure 5–92 Apply felt strips.

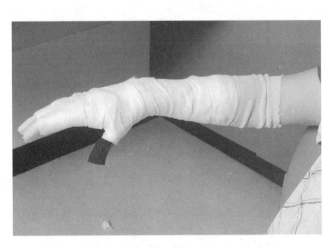

Figure 5–93 Apply padding three fourths the length of the forearm, past DIP joints. The padding is finished.

Method A

Aim middle roll of padding to middle of web space. Tear padding vertically at the third or fourth metacarpal on the volar surface. Pull padding up through web space. Tear vertically again on the dorsal surface at the second metacarpal. Wrap ends back around the MCP joint of the thumb. Wrap back around CMC joints ulnar to radial on the volar surface, radial to ulnar on the dorsal surface. Repeat.

Method B

Aim middle roll of padding to middle of web space. As padding is pulled up through web space, tear horizontally at the third metacarpal on the dorsal surface. Wrap proximal half of padding around metacarpal of the thumb and lay the other half of padding across the MCP joints. Wrap back around the CMC joint ulnar to radial on the volar surface, radial to ulnar on the dorsal surface. Repeat.

5. Apply plaster ½ inch below top of padding at proximal end of forearm distally over the MCP joints of the hand. Follow MCP wrist fabrication procedure. Fashion a figure-eight around the thumb, ulnar to radial on the volar surface of the hand. Do not squeeze the plaster through the web space. Lay plaster at the web space so that it lies partially in the trough formed by the padding and over the MCP joints. Angle the plaster and tuck to continue unrolling radial to ulnar on the dorsum of the hand and back around the volar surface of the CMC joint. Come through the web space with the plaster only three times, otherwise it may become too bulky.

6. Form palmar arch. Apply pressure on the volar surface with index, middle, and ring fingers and counterpressure at the wrist using the volar surface of your hand (Figure 5–94). Do not apply counterpressure with thumb or fingertips. Apply additional layers of plaster to reinforce cast if needed.

7. Apply the three sets of strips. Apply first set on the center volar surface of the hand from fingertips to midforearm (Figure 5–95). Apply second set slightly lateral to the first set (Figure 5–96). Apply third set slightly medial to the first set (Figure 5–97).

8. Reinforce with an additional roll of plaster applied in a circular manner, similar to the wrist cast, covering the strips from proximal to the MCP joints (Figure 5–98).

9. Cut padding and stockinette over the dorsal surface, as described in the drop-out elbow cast procedure. Cut distally down the center of the padding on the dorsal surface, gliding scissors along the stockinette (Figure 5–99). Stop at the PIP joint of the middle finger and angle cut toward the little finger and index fingers, making a triangle (Figures 5–100 and 5–101).

10. Cut through stockinette in the same way as described above. Cut away excess padding, leaving ½ inch exposed beyond plaster edge. Pull stockinette over the edge of plaster and fix in place with plaster strips (Figure 5–102 [dorsal view], Figure 5–103 [volar view]).

THUMB ENCLOSED CAST

Rationale for Use

The position of the wrist influences the ability of the thumb to participate in prehension and manipulation of objects. When the wrist is pulled into flexion and ulnar deviation, for example, the thumb becomes biomechanically adducted and develops a shortened web space. This position prevents the thumb from achieving a true opposition pattern needed for effective pinch, grasp, and release. The improved muscle balance and joint alignment gained through a combination wrist and thumb casting program may lead to improved thumb function.

The degree of thumb involvement varies greatly within the patient population. The most common deformity, thumb-in-palm, generally is caused by an abnormal muscle pull of the flexor pollicis longus or brevis, adductor pollicis, and/or the first dorsal interossei, which in turn results in a fixed myostatic contracture.

The wrist cast with the thumb enclosed may be used to improve palmar expansion gradually. Unless a tightly flexed thumb is sufficiently realigned, it will block the development of the transverse or carpal arch, as well as the oblique arch that links the thumb to the fingers for opposition.

The thumb portion of the cast should be molded comfortably in the web space, stretching across the thenar eminence. If the client has a tightly contracted thumb, hyperextension at the interphalangeal (IP) or MCP joint of the thumb may occur. Thus it is important to contour the casting material carefully to the thenar eminence while supporting the MCP and IP joints. Sufficient tension at submaximal range should be used to elongate thumb abductors and extensors gradually, with the wrist positioned in neutral or in 10° to 15° of extension.

A series of wrist casts that enclose the thumb can be used to elongate the shortened thumb flexors and adductor muscles gradually. This will help promote the rebalancing and control of the opposing muscle groups. However, follow-up after casting must include strengthening and a muscle reeducation program to optimize thumb alignment and function.

The combination wrist and thumb cast can be especially helpful when treating a patient who presents with

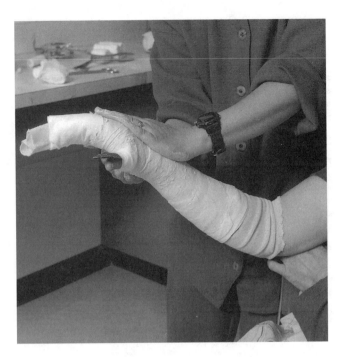

Figure 5–94 Form palmar arch.

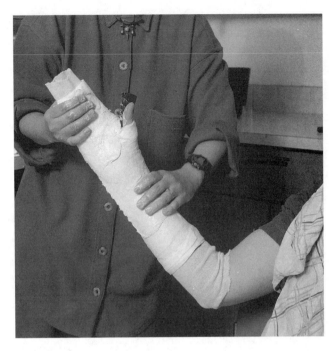

Figure 5–95 Apply first set of strips on center volar surface.

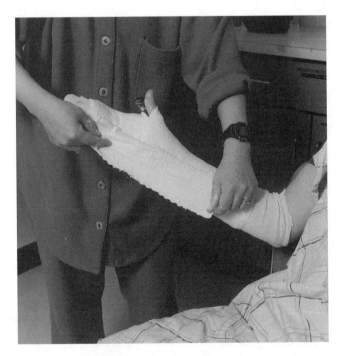

Figure 5–96 Apply second set of strips slightly lateral to first set.

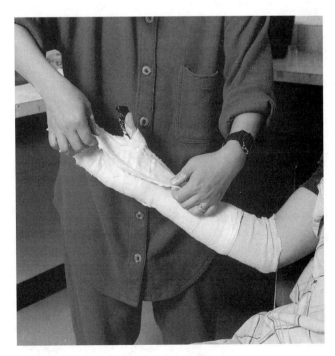

Figure 5–97 Apply third set of strips slightly medial to first set.

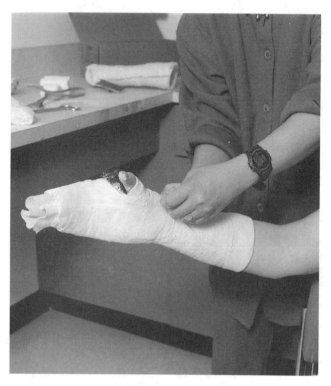

Figure 5–98 Reinforce with additional plaster roll.

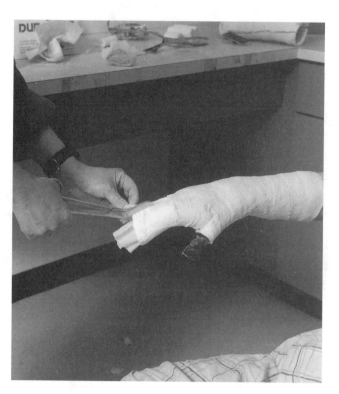

Figure 5–99 Cut padding over dorsal surface.

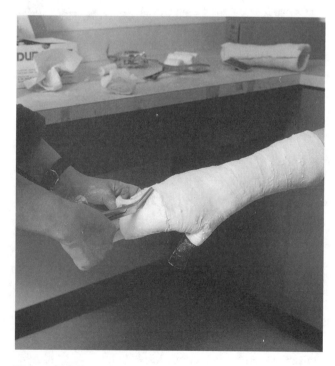

Figure 5–100 Angle cut.

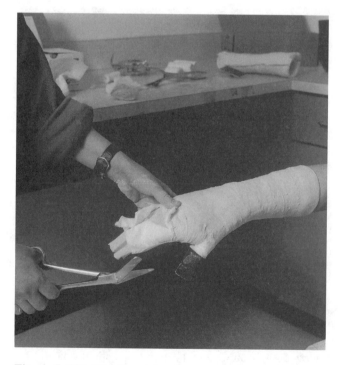

Figure 5–101 Making a triangle.

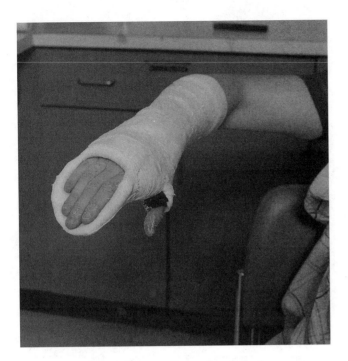

Figure 5–102 Dorsal view of completed platform.

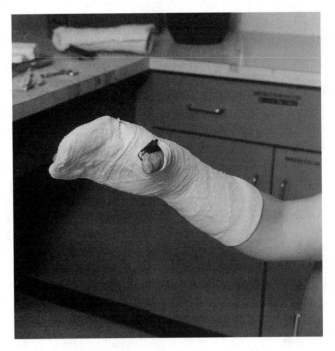

Figure 5–103 Volar view of completed platform.

increased tone in the hand and arm. By positioning the thumb in extension or abduction, a relaxing effect can be gained in the tightness in the fingers, wrist, or arm. Through evaluating the results of palpation, the therapist will choose the most relaxing position in which to cast the thumb, again observing submaximal range. The position of the thumb promotes an overall inhibition of the abnormal tone through the gentle stretch and pressure placed on the thenar eminence.

Required Casting Materials

Prepare materials and apply the cast according to the procedures for the specific cast being fabricated, such as long arm cast, platform cast, wrist cast, and MCP cast.

Cast Application Instructions

1. Apply stockinette and felt strips as described for the specific cast being fabricated.
2. Apply padding as described for the specific cast being fabricated.

3. Wrap padding up to the web space and wrap two to three times around the MCP joint past the IP joint of the thumb (Figure 5–104). If the padding starts to bulk, tear the padding distally to make the material conform smoothly to the thumb.
4. Continue to figure-eight wrap around the thumb (Figure 5–105). As described in the wrist cast procedure, use either of the following methods:
   Method A
   Aim middle roll of padding to middle of web space. Tear padding vertically at the third or fourth metacarpal on the volar surface. Pull padding up through web space. Tear vertically again on dorsal surface at the second metacarpal. Wrap ends back around the MCP joint of the thumb. Wrap back around CMC joints ulnar to radial on the volar surface, radial to ulnar on the dorsal surface. Repeat.
   Method B
   Aim middle roll of padding to middle of web space. As padding is pulled up through the web space, tear horizontally at the third metacarpal on the dorsal surface. Wrap proximal half of pad-

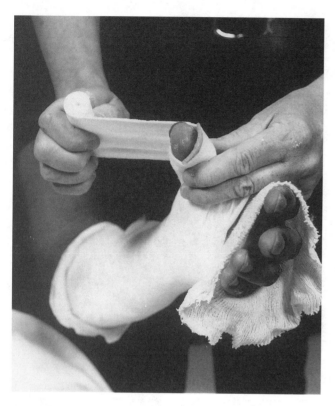

Figure 5–104 Wrap around MCP joint past IP joint of the thumb.

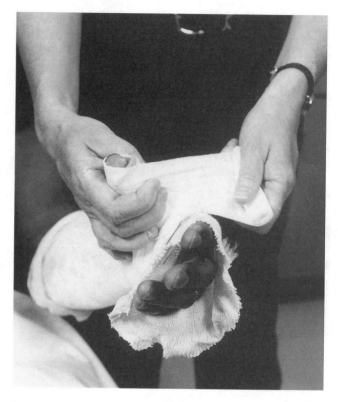

Figure 5–105 Continue to figure-eight wrap around the thumb.

ding around metacarpal of thumb and lay the other half of padding across the MCP joints. Wrap back around the CMC joint ulnar to radial on the volar surface, radial to ulnar on the dorsal surface. Repeat.

5. Apply the plaster. Do not squeeze the plaster together at the web space. Lay the plaster between the web space and slightly toward the MCP joint of the thumb (Figure 5–106). During the second time through the web space, wrap the plaster two to three times up to the middle of the nailbed of the thumb (Figure 5–107).

6. After applying one to two rolls of plaster, form a palmar arch and carefully contour the thenar eminence while supporting the MCP and IP joints of the thumb (Figure 5–108). Apply pressure on the volar surface with index, middle, and ring fingers and counterpressure at the wrist, using the volar surface of your hand. Do not apply counter-

pressure with thumb or fingertips. Apply additional layers of plaster to reinforce cast if needed.

7. Turn stockinette back over plaster edge and secure with plaster strips.

8. Wash hands and flare proximal and distal edges with a circumferential motion combined with outward pull of the index finger.

9 Trim the padding at the distal end, allowing the tip of the thumb to be seen for skin monitoring and check (Figure 5–109).

METACARPOPHALANGEAL WRIST CAST

Rationale for Use

When spasticity of the intrinsics causes flexion contractures of the MCP joints, this prevents the patient from fully extending at these joints. The patient can only flex and extend the fingers with the MCP joints in

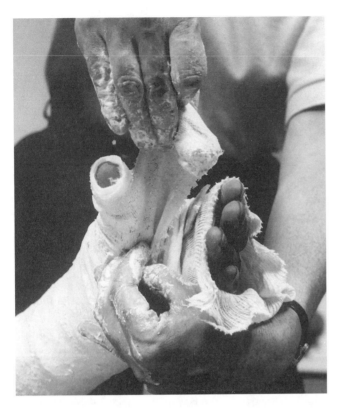

Figure 5–106 Lay plaster between web space toward the MCP joint of the thumb.

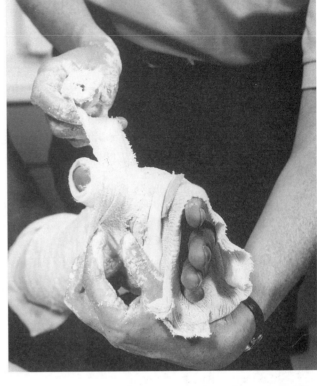

Figure 5–107 Wrap plaster two to three times around the thumb.

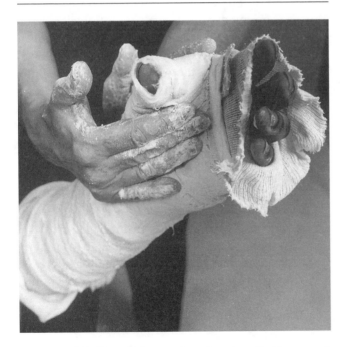

Figure 5–108 Form palmar arch and contour the thenar eminence.

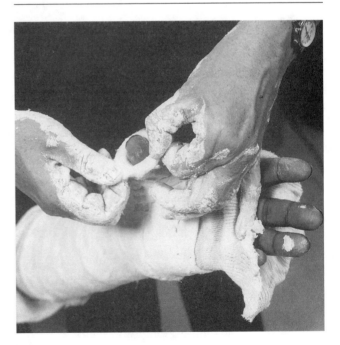

Figure 5–109 Trim padding at the thumb.

flexion. If untreated, this can lead to an intrinsic-plus type hand.

The MCP wrist cast is similar to a rigid circular wrist cast; however, it also encloses the MCP joint. When applying this cast, follow the same procedures as for the rigid wrist cast, placing minimal stretch on the joints and muscles and maintaining the wrist-hand in submaximal range. The cast should be molded to support the distal transverse arch located at the distal row of the metacarpal bones and the hand placed in the functional position.

This cast is left in place for 5 to 7 days. The patient can actively work on finger flexion and extension of the PIP and DIP joints while positioned in the cast. This cast can gradually elongate the intrinsic musculature to facilitate active extension of the interphalangeal joints when the MCP joints are extended.

The materials and cast application procedures are similar to those for the rigid circular wrist cast.

Required Casting Materials

- Stockinette: 3-inch width for most adults
- Cast padding: 3-inch width, three to five rolls depending on size of arm
- Plaster: 3-inch width, four to five rolls depending on size of arm
- Felt: four strips
  one 1½-inch width by circumference of forearm below elbow joint
  one 1½-inch width by circumference of forearm over ulnar styloid
  one 1½-inch width by 5 inches
  one 1-inch width by 3 inches
- Paper tape
- Towels, gown to cover patient
- Water: quite warm but not hot to touch
- Bandage scissors

Cast Application Instructions

1. Measure stockinette from the olecranon to the DIP joints (Figure 5–110), and approximate placement of the thumb. Cut a straight slit (¼ inch) for the thumb. Apply stockinette.
2. Apply felt strips as described for the rigid wrist cast.

   a. distal to the olecranon
   b. over the ulnar styloid
   Use the two small strips for the thumb piece. Fold larger strip in half and cut a heart shape one third from the top. Place the smaller felt strip through the hole of the larger strip. Place the felt strip piece through the thumb and along the radial border of the forearm with the second piece covering the web space. Tape felt in place.

3. Apply padding. Use same procedure as for rigid wrist cast; however, extend distally to the PIP joint (Figure 5–111). It is easier to begin distally at the wrist because the circumference of the forearm is smaller. Wrap circumferentially, overlapping 1½ inch on each wrap. If bulking occurs at the narrow end while wrapping circumferentially, hold padding close to forearm and tear at the opposite end to contour to the shape of the forearm. The padding extends fully three fourths the length of the forearm distally to the PIPs. Apply four to five layers at both proximal and distal ends.

4. When wrapping distally at the hand, figure-eight wrap around the thumb. As described in the wrist cast procedure, use either of the following methods:

   Method A

   Aim middle roll of padding to middle of web space. Tear padding vertically at the third or fourth metacarpal on the volar surface. Pull padding up through the web space. Tear vertically again on the dorsal surface at the second metacarpal. Wrap ends back around the MCP joint of the thumb. Wrap back around CMC joints ulnar to radial on the volar surface, radial to ulnar on the dorsal surface. Repeat.

   Method B

   Aim middle roll of padding to middle of web space. As padding is pulled up through the web space, tear horizontally at the third metacarpal on dorsal surface. Wrap proximal half of padding around the metacarpal of the thumb and lay the other half of padding across the MCP joints. Wrap back around CMC joint ulnar to radial on the volar surface, radial to ulnar on the dorsal surface. Repeat.

5. Apply plaster ½ inch below the top of padding at the proximal end of the forearm and unroll to the

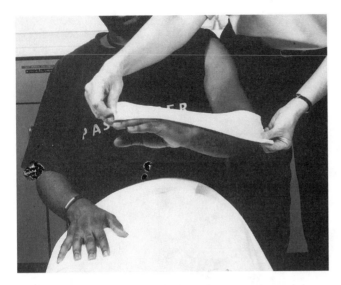

Figure 5–110 Measure stockinette from the olecranon to the DIP joints.

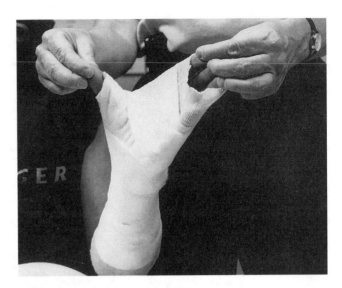

Figure 5–111 Padding the PIP joint.

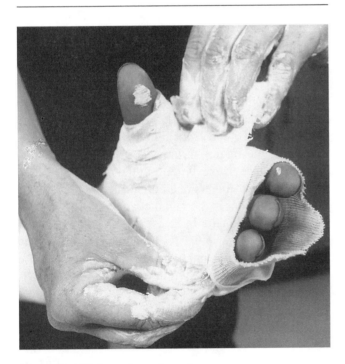

Figure 5–112 Lay plaster in web space.

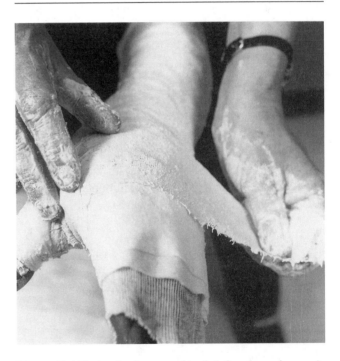

Figure 5–113 Angle plaster and tuck (after the web space).

distal end past the MCP joints, leaving ½ inch of padding distally. Apply figure-eight wrap around the thumb, ulnar to radial of the volar surface of the hand. Do not squeeze the plaster through the web space. Lay the plaster at the web space so that it lies partially in the trough formed by the padding and past the MCP joints (Figure 5–112). Angle the plaster and tuck to continue unrolling radial to ulnar on the dorsum of the hand (Figure 5–113) and back around the volar surface of the CMC joint.

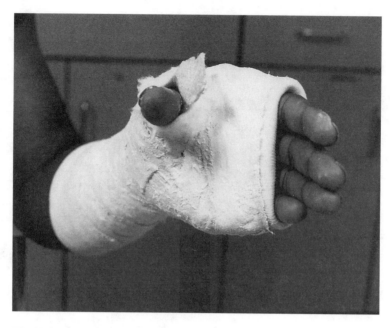

Figure 5–114 Completed MCP wrist cast.

Come through the web space with the plaster only three times, otherwise it may become too bulky.

6. After applying one to two rolls of plaster, form a palmar arch as described for the rigid wrist cast. Apply pressure on the volar surface with index, middle, and ring fingers and counterpressure at the wrist using the volar surface of your hand. Do not apply counterpressure with thumb or fingertips. Apply additional layers of plaster to reinforce cast if needed.

7. Wash hands and flare proximal and distal edges with circumferential motion combined with outward pull of the index finger. Flare the MCP area of the thumb. Do not apply counterpressure with thumb. To check for tightness at the distal end, shift the fingers in the cast in a radial direction and place index finger into the ulnar edge of the cast. Shift fingers in an ulnar direction and place index finger in the radial edge of cast.

8. Turn stockinette back over plaster edge and secure with plaster strips. Finish the thumb area by folding the felt and secure with the stockinette for the web space. Completed MCP wrist cast is shown in Figure 5–114.

CHAPTER 6

Lower Extremity Assessment

Paula Goga-Eppenstein and Terry Murphy Seifert

Before starting lower extremity casting, considerations include the following: efficacy and appropriate usage of this intervention, a thorough pre- and postintervention assessment, and interim assessments during the casting series. Components of the assessment include passive range of motion, muscle length, spasticity, sensation, active movement and functional use, and position of the limb at rest.

DECIDING TO CAST

The main reasons for use of casting are passive loss of motion, restricted active movement due to soft tissue tightness, and increased muscle tone. When any of these factors are present, an assessment is essential to determine the probable effectiveness of this treatment intervention. Evaluation of x-ray results and consultation with the physician are necessary to rule out the presence of any orthopaedic impairments that would not be affected by casting. For instance, in the case of heterotopic ossification, mobilization of the joints is recommended to prevent further stiffness and loss of motion.

Another factor to consider before initiating casting is the patient's response to previous interventions. A patient who responds well to passive stretching and has no significant x-ray findings, but the gains are lost from session to session, might be considered for casting. A patient who has already received a series of splints and casts may not respond as favorably with this intervention. Further considerations for use of casting are discussed later in this chapter.

Assessment of potential functional gains and patient goals is essential in determining whether casting should be considered. In the presence of loss of motion, casting may result in the patient's being able to manage his or her lower extremity more easily, allow for ease in performing self-care/hygiene activities, enhance weight bearing in sitting or stance, or prevent skin breakdown.

ASSESSMENT

Once the determination is made that casting is appropriate, specific evaluations are performed to determine and document initial baseline measures and progress achieved. These measures are taken for areas that may be affected by casting before the first cast application and upon each cast's removal. Other evaluation tools may be used to address specific patient goals. For example, gait or posture analysis may be used to assess whether gains in joint range of motion have carried over into function. Standard evaluation items are given below.

Passive Range of Motion

Goniometric measurements of the entire extremity should be recorded. In muscles that cross more than one joint, it is possible to see changes in range in adjacent joints. For example, with a serial knee cast in place, increased range in knee extension might be noted upon cast removal; however, ankle dorsiflexion may be decreased. This decrease may be due to the gastrocnemius muscle's pulling the ankle into plantarflexion as it is stretched across the knee. Measurements are made both pre- and postcasting. Periodic measurements are taken following each cast's removal to monitor any loss or additional gains in range of motion.

Muscle Length

Goniometric measurements of joints crossed by multijoint muscles can be used to evaluate muscle length. In doing this, only one joint is allowed to vary at a time. Ankle range can be measured with the knee in full extension or 90° of flexion. When comparing initial measurements with subsequent measurements, the "fixed" joint position must be the same.

Soft Tissue versus Joint Limitation

When evaluating passive range of motion and muscle length, the relation between muscle contracture and joint limitations must be considered. Assessing joint play through glides may differentiate the origin of the problem. If the problem is more mechanical in terms of restricted joint play, performing glides/mobilization may loosen up any restrictions and allow for further increases in range of motion. Since casting appears to have more of an effect on changes in muscle, joint mobilization may address any limitations imposed by tight ligaments.

Sensation

Evaluation of sensation is completed pre- and postcasting. A patient with impaired sensation may not be able to report any problems that are due to constriction, pressure, or rubbing. Care in cast application and more frequent monitoring of visible skin areas may be required. If sensory status changes, possibly due to swelling or impaired circulation while the cast is on, the cast should be removed and the underlying area evaluated for any adverse reactions.

Spasticity or Abnormal Muscle Tone

The ability to measure muscle tone reliably in the clinical setting is a challenge. One can evaluate muscle tone by measuring the joint angle at which resistance is encountered following a quick stretch. While using this method, it is difficult to control for variances in velocity and force in subsequent measurements. The Ashworth Scale classifies spasticity as mild, moderate, or severe when a quick stretch is elicited.

Skin Integrity

Observations about the limb's skin surface should be documented. Any preexisting abrasions should be noted indicating size, location, and color.

Circulation

Assess for pulse, skin blanching, and nailbed refill. Any tendency for edema should be noted and measured. Temperature of the limb to be casted should be compared with that of the noncast limb. Any discoloration should be noted and monitored.

Motor Control

The ability to perform specific movements and the speed at which they can be performed are evaluated. For instance, evaluate the patient's ability to plantar- and dorsiflex the ankle with the knee first in extension, then in flexion. Speed of motion can be assessed by counting the number of times a movement can be performed in a specific amount of time. Control of movement can be evaluated by assessing how coordinated and fluid the motion is.

Functional Use

This assessment focuses on the ability to incorporate use of the extremity during activities of daily living, for instance, the ability of the foot to maintain full ground contact during standing activities or the ability to have sufficient dorsiflexion for toe clearance during gait.

Note: While all of the above assessments are recommended prior to initiating a casting program, there may be instances when all are not performed. For example, a patient with low cognitive function and little active motion may not require evaluation of motor control and functional use.

ASSESSMENT BETWEEN SERIAL CASTS

While attention to all of the areas is important in assessing the effectiveness of a casting program, casts should be removed and replaced as quickly as possible to maintain the gains achieved from one cast to the next. Keeping this in mind, assessment between casts should focus on ensuring that gains are being made in several areas.

Range of Motion

Between casts, passive range of joint motion is measured and compared with prior measurements. If no change is evident following removal of the two successive casts, continuation should be questioned and other treatments should be considered. Stiffness and loss of motion in the opposite direction need to be assessed. Some stiffness is to be expected after cast removal. The limb should be ranged actively, if possible, and passively to counteract the stiffness and preserve full available motion. Casting is terminated when full range of motion is achieved or there is no change in motion following two cast applications.

Skin Integrity

Observe skin condition with attention to blisters and red or open areas resulting from the cast. If present, red areas should dissipate in 5 to 10 minutes. Areas that do not resolve in 10 minutes and blisters or open areas require a decision to be made about whether casting should continue or be terminated. If there is actual skin breakdown, special dressings may be used and casted over only under physician direction. In some cases, the cast program may need to be put on hold temporarily until the area resolves.

Circulation

The presence of edema is noted. If this condition has increased or is now present, the cast program may be put on hold temporarily depending upon severity.

Spasticity

Goniometric measurement of joint position when the stretch reflex is initiated is used as an indicator for change. If there is no change, but range of motion is improved, casting is continued.

Functional Use/Motor Control

A brief assessment might be performed.

Summary

Assessment is essential in evaluating and monitoring the effectiveness of a casting program because it provides a basis for deciding whether to continue or terminate use of this intervention. Therefore, proper assessment is key in the effective use of lower extremity casting.

USE OF CASTS TO MANAGE LOWER EXTREMITY CONTRACTURES

Serial Casting

Serial casting involves a series of progressive, corrective casts that are applied in succession. With each cast, the limb and underlying soft tissue structures are gradually stretched. By using a low-load, prolonged stretch, soft tissue trauma is minimized and the patient is able to better tolerate the new position. The cast is then left in place for 5 to 7 days. This practice is consistent with prior casting studies such as that by Booth et al.[1] Between casts, not only is the skin checked and cleaned, but the limb is mobilized to prevent stiffness from prolonged immobilization. Range of motion gains are measured and documented. If no changes occur after two casts, serial casting is terminated and a resting bivalve splint is fabricated and used to prevent further loss of motion.

Considerations for Use of Serial Casting

When trying to determine whether the patient would benefit from serial casting, the following should be considered:

- Certain functional limitations justify serial casting. They include range of motion limitations at the ankle and/or knee secondary to soft tissue contractures that interfere with distal weight bearing at the foot during transfers or standing activities, wheelchair positioning, or weight bearing at the foot for normal postural alignment.

- Serial casting is most effective within the first 6 months following the injury.[1] The greater the length of time since the onset of the injury, the less chance of fully correcting the contracture.

- Casting is less successful when prior attempts have been made to correct for the deformity, through either surgery or casting. Especially in cases of surgical intervention, scar formation may result in additional shortening of soft tissue.

- In the presence of fixed orthopaedic contractures (e.g., heterotopic ossification), casting will not result in improved range of motion.

- In the presence of dystonic movements or fluctuating levels of muscle tone, especially spasms, care must be taken during the application procedure to prevent skin problems from developing. If the cast is poorly tolerated (e.g., the patient complains of pain), remove the cast and check the skin. If no skin problems exist, another cast may be applied.

- Consideration of the patient's ability to tolerate a cast for a period of time is important. Those individuals who respond best are those who tolerate passive mobilization of their joints without adverse behavior such as agitation or crying. Patients who understand the rationale for this procedure and cooperate once the cast is applied will tend to tolerate having the cast on for 1 week.

- Adequate circulation should be present. Check pedal pulses and assess nailbed refill or skin blanching. In the presence of fluctuating levels of edema, casting should not be performed because of the risk of complications.

- Skin integrity must be assessed. Sensitivity of the skin to any pressure or friction from the cast may require more care during the application to ensure a good mold. Any minor abrasion or potential area of skin breakdown must be assessed. Even if present, the joint could be casted, but extra care must be taken with padding and during the application.

Indications for Serial Short Leg/Long Leg Casts

The following indications have been compiled from the literature and from clinical experience:

- Soft tissue contracture of the hindfoot and/or forefoot prevents normal excursion at the ankle and forefoot. The most common foot and ankle deformity following an insult is equinus. An equinus deformity consists of supination, plantarflexion, adduction, and inversion. Casting reduces this deformity through dorsiflexion at the ankle, abduction of the midtarsal joint, and eversion of the subtalar joint.

- Soft tissue contractures of the knee limit flexion or extension of the joint. The most common deformity is a flexion contracture. Loss of motion can result from shortening of the gastrocnemius muscle, or in combination with the hamstrings. While casting of both the knee and ankle in combination is possible, it is not recommended initially secondary to the increased distribution of pressure and the potential for skin breakdown and changes in muscle tone. Instead, start by casting the ankle first for a few casts to increase soft tissue length before incorporating the knee.

- There is poor carryover of results from passive manual stretching and/or joint mobilization. Through sustained continuous stretch, the plastic components of connective tissue are favored (see Chapter 1).

- Onset is fairly recent. Results obtained with casting may not be as significant in contractures older than 1 year. This is more true for adults than children.

- There is mild to moderate hypertonus. Casting will have minimal results in the presence of severe tone, and other options should be considered (see Chapter 10).

- There is good skin integrity.

- There is adequate circulation.

Contraindications for Serial Short/Long Leg Casts

Serial casting may be harmful or ineffective if full consideration is not given to all the pros and cons. The following should be considered prior to the start of serial casting:

- The contractures are long-standing, greater than 1 to 2 years in adults.
- The presence of any severe orthopaedic abnormality would restrict stretching of soft tissue structures (e.g., surgical fixation, heterotopic ossification) or require special consideration/techniques beyond the scope of practice, such as setting fractures or reducing subluxations.
- The patient is extremely agitated and at risk for a self-inflicted injury from the cast.
- An uncooperative patient and/or family may not comply with follow-up monitoring and/or recommendations.
- In the presence of open skin areas, grade II or worse, casting is contraindicated, especially if frequent dressing changes are needed.
- In the presence of severe and/or fluctuating levels of edema, constriction and/or compression of underlying tissues and nerves can result.
- The patient has severe tone.

Dynamic Weight-Bearing Casts

As the name implies, this is a weight-bearing cast, so full joint range of motion is optimal. A series of casts may be necessary to reach this point. Dynamic weight-bearing casts incorporate the use of a contoured footboard into the application process. The footboard serves to hold the foot in a position of partial correction of the deformity.

The contoured footboard consists of built-up and depressed areas to reduce sensory input to certain areas of the foot along with supporting other spots. The depressed area beneath the calcaneous allows for the heel to remain centered or move medially or laterally to align the hindfoot. Plaster built up under the medial longitudinal arch or peroneal arch will aid in correcting excessive pronation or supination. Plaster built up under the toes, especially the second through fifth, will support the toes in extension and reduce toe clawing.

The basic premise in using a footboard is to maintain the foot in good alignment in a decreased hypertonic state for a period of time. This frees up the therapist to work on establishing more normal movement patterns. The primary concerns in footboard fabrication are to provide for good heel alignment, to correct for deformities, and to reduce contact in the reflexogenic areas of the foot. When considering fabrication of a footboard, keep in mind that this technique is demanding and requires a significant time commitment. Also, fabrication should be done by experienced personnel due to the concern about the patient's developing pressure sores if the cast is poorly constructed.

Indications for Dynamic Weight-Bearing Casts

The following should be considered prior to fabrication:

- Optimally, the patient should possess full range of motion at the ankle joint since this is a weight-bearing cast.
- Increased stability at the knee, ankle, and foot is provided by controlling for varus/valgus movement at the ankle, pronation/supination at the forefoot, clawing at the toes, and hyperextension at the knee.
- By depressing certain areas in the fabricated footboard, sensory input that results in pathological posturing is reduced.
- Pressure of generalized hypertonus can be reduced through handling techniques.

Contraindications for Dynamic Weight-Bearing Casts

Contraindications for dynamic weight-bearing casts are the same as those for short/long leg casts.

REFERENCE

1. Booth BJ, Doyle M, Montgomery J. Serial casting for the management of spasticity in the head-injured adult. *Phys Ther.* 1983;63:1960–1966.

CHAPTER 7

Types of Lower Extremity Casts

Paula Goga-Eppenstein and Terry Murphy Seifert

SHORT LEG SERIAL CAST—VARIATION I

Rationale

The short leg serial cast is the basic cast used to increase ankle motion. It is usually applied with the patient in sitting position. Typically, the cast is left in place for 5 to 7 days. This cast may also be applied in supine or prone position. The supine position is used for patients unable to tolerate sitting during the casting procedure. The hip and knee should be held in flexion to minimize the influence of extensor tone. The prone position may be used for patients who tolerate this position. This is the easiest position to maintain knee flexion. The perspective for foot alignment may be challenging because of the reverse view, however.

Prepare Materials

- Stockinette: 2- or 3-inch depending on size of lower extremity
- Felt padding for bony prominences
- Cotton webril: 2-, 3-, or 4-inch depending on size of lower extremity, three to four rolls
- Plaster or fiberglass: 2-, 3-, or 4-inch, approximately three or four rolls
- Paper tape

- Warm water
- Bandage scissors
- Towels
- Gown to cover patient

Fabrication Procedure

1. Position the patient. Drape him or her with sheets and towels.
2. Measure stockinette starting from the fibular head extending 2 inches past the toes. Cut two lengths of stockinette.
3. Apply the stockinette to the lower extremity one layer at a time. Smooth out all wrinkles (Figure 7–1). If unable to smooth out wrinkles at the anterior surface of the joint, cut a small dart and overlap the layers of stockinette (Figures 7–2 and 7–3).
4. Cut felt pads to cover bony prominences (especially malleoli, navicular heads, if prominent, and metatarsal heads). Secure pads in place with tape (Figure 7–4).
5. Position the ankle at submaximal range.
6. Begin wrapping with webril. Start distally and work proximally (Figure 7–5). Keeping the webril close to the extremity, wrap circumferentially, rolling it smooth to avoid wrinkles. Overlap the webbing by half its width.

All areas should have two to three layers of webbing. The webbing should be smooth, but taut. Tear the webbing at any point to smooth out wrinkles. Apply approximately three layers of webril (Figure 7–6).

7. Grasp the end of the plaster roll and dip in warm water for approximately 2 or 3 seconds (Figure 7–7). Squeeze out excess water.

8. Holding the roll, begin wrapping with the plaster. Start distally to set the ankle. Leave ½ inch of webril exposed at the toes. Wrap with a figure-eight around the ankle to secure the joint. Smooth the plaster with the palms of your hands as you wrap because using the fingers to smooth will lead to indentations (Figures 7–8 and 7–9).

9. Repeat the procedure with the remaining plaster rolls as you wrap upward along the leg. Leave ½ inch of webril showing below the fibular head. A total of three layers of plaster should be used. Smooth any wrinkles or ridges. You may need to moisten your hands with water if the plaster begins to dry (Figure 7–10).

10. Fold back stockinette at the knee and toes. All five toes should be visible. Secure stockinette with a roll of plaster (Figures 7–11, 7–12, 7–13, 7–14, and 7–15).

11. The plaster will set in 15 to 20 minutes but will not harden completely for 24 hours.

12. Assess skin color, temperature, and sensation.

13. Serial casts remain on for 5 to 7 days.

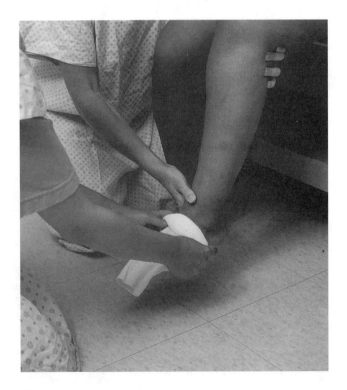

Figure 7–1 Apply stockinette one layer at a time.

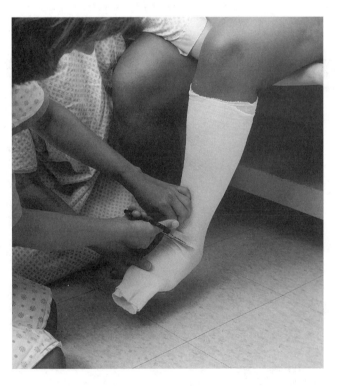

Figure 7–2 If unable to smooth wrinkles, cut a small dart.

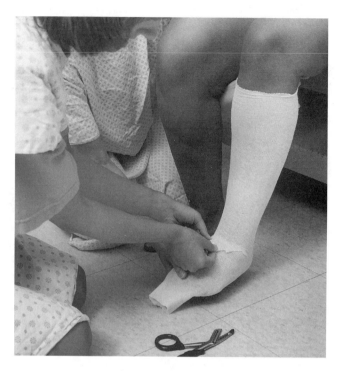

Figure 7–3 Overlap cut layers of stockinette.

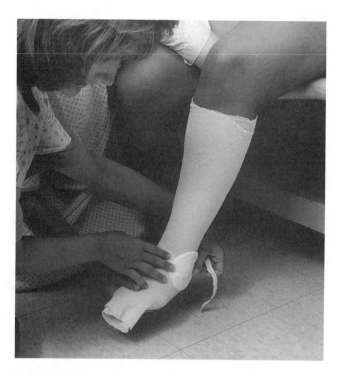

Figure 7–4 Cover bony prominences with felt pads.

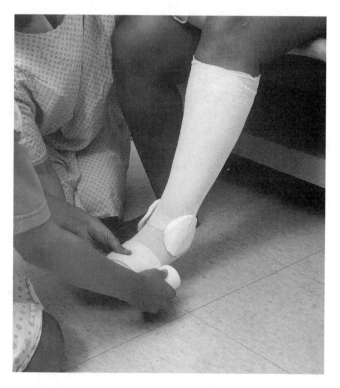

Figure 7–5 Begin wrapping webril distally, work proximally.

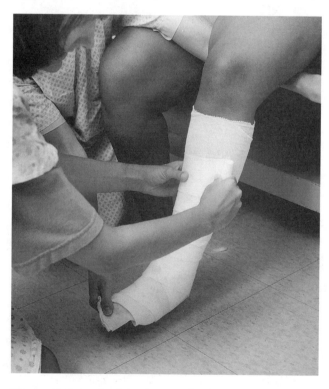

Figure 7–6 Apply three layers of webril, smoothing each layer.

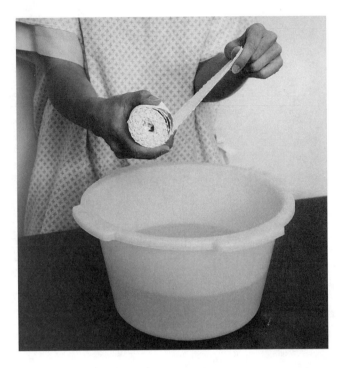

Figure 7–7 Dip plaster in warm water.

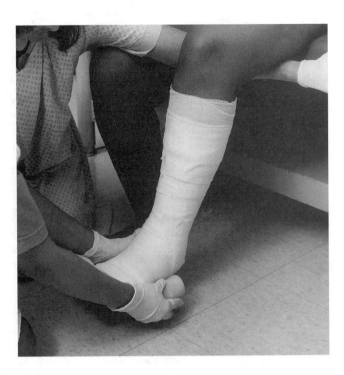

Figure 7–8 Wrap plaster, begin distally at toes.

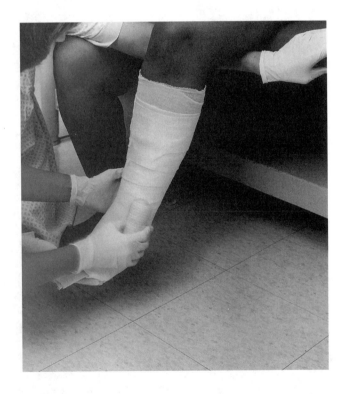

Figure 7–9 Smooth layers with your palms.

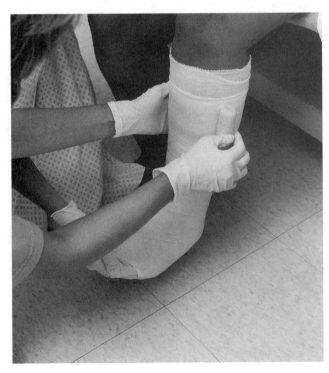

Figure 7–10 Continue wrapping proximally.

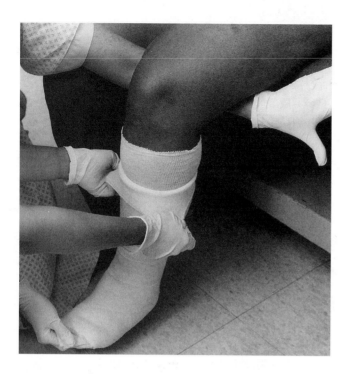

Figure 7–11 Fold back first layer of stockinette.

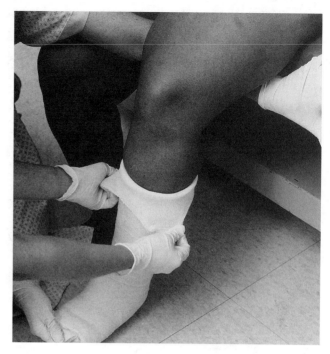

Figure 7–12 Fold back second layer of stockinette.

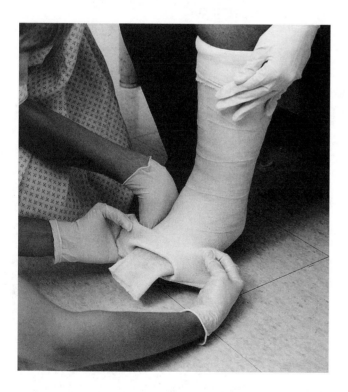

Figure 7–13 Fold back first layer of stockinette.

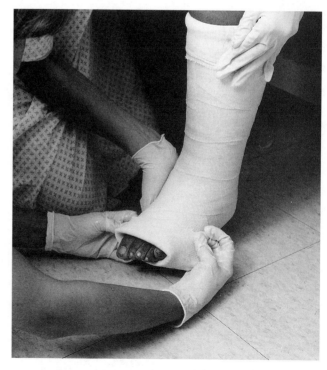

Figure 7–14 Fold back second layer of stockinette.

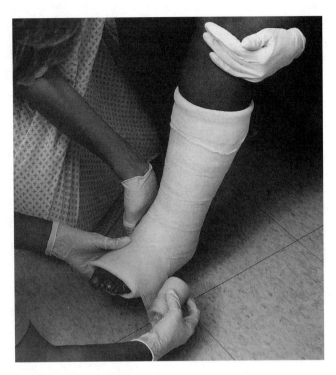

Figure 7–15 Secure stockinette with layer of plaster.

SHORT LEG SERIAL CAST—VARIATION II (SLIPPER MOLD)

Rationale

- The slipper maintains forefoot alignment during the casting procedure.
- The slipper would *not* be used when there is a need to control the toes.
- The slipper procedure allows for molding and contouring of the ankle.

Prepare Materials

- Stockinette: 2- to 3-inch depending on size of lower extremity
- Felt padding for bony prominences
- Cotton webril: 2-, 3-, or 4-inch depending on the size of the lower extremity, three to four rolls
- Plaster rolls: 2-, 3-, or 4-inch, approximately three to four rolls
- Five-inch plaster splints: 10 splints
- Paper tape

- Bandage scissors
- Towels, sheets, gown for patient

Fabrication Procedure

1. Follow steps 1 through 6 of Short Leg Serial Cast—Variation I.
2. Mold a plaster slipper around the foot to hold the subtalar joint/metatarsal joint in a neutral position. Taking 5 plaster splints, dip in warm water for approximately 2 or 3 seconds. Squeeze out excess water. Apply the five plaster splints starting from the head of the first metatarsal, extend around the back of the heel, to the head of the fifth metatarsal (Figure 7–16). Overlap the plaster splints on the plantar surface. Mold the splints under the arch and around the heel. Wrap the remaining five plaster splints around the forefoot. Hold this position until the plaster sets (Figures 7–17A and 7–17B).
3. If a footplate is not being used, place five plaster splints along the plantar surface (Figures 7–18 and 7–19). Remember to support toes.
4. Follow steps 7 through 12 of Short Leg Serial Cast—Variation I.

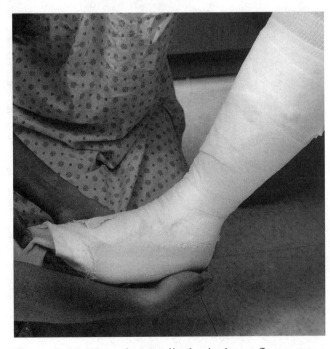

Figure 7–16 Wrap plaster splint beginning at first metatarsal, around heel to fifth metatarsal.

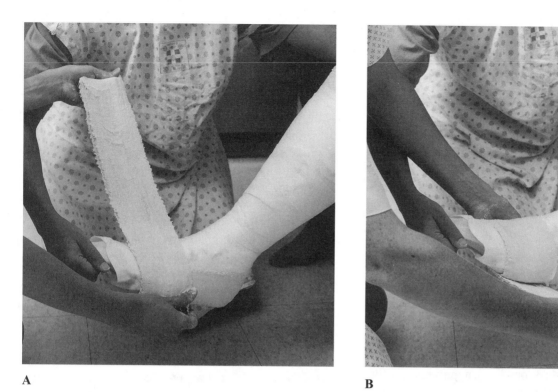

A   **B**

Figure 7–17 (A) Wrap plaster splints around forefoot; (B) smooth in the direction you have wrapped.

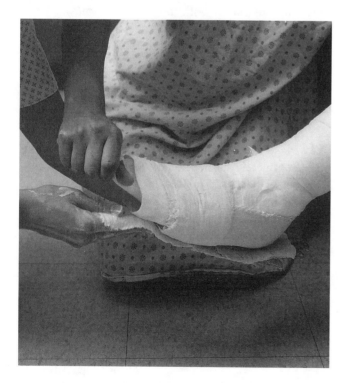

Figure 7–18 Apply plaster splint along plantar surface.

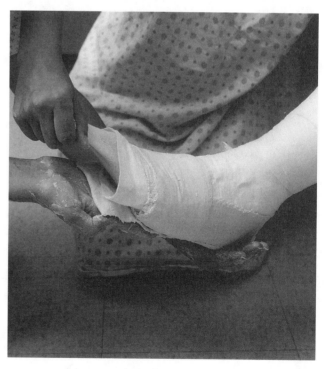

Figure 7–19 Smooth plaster along plantar surface of foot.

SERIAL KNEE CAST

Rationale

The serial knee cast is the basic cast used to improve knee range of motion. It is typically applied with the patient in supine position. This cast is also left in place for 5 to 7 days. Care must be taken that the cast extends far enough above and below the knee joint to stabilize the position.

Prepare Materials

- Stockinette: 3- or 4-inch depending on size of extremity
- Felt padding for bony prominences
- Tape
- Cotton webril: 3- or 4-inch depending on size of lower extremity, five to six rolls
- Plaster/fiberglass: 3- or 4-inch, approximately five to six rolls
- Warm water
- Gloves (if using fiberglass)
- Towels and gown to cover patient
- Bandage scissors

Fabrication Procedure

1. Put patient in supine position. Drape the patient and area with sheets and towels.
2. Measure stockinette starting from 1 inch below the pubis extending to just above the malleoli. Cut two lengths of stockinette.
3. Apply the stockinette to the limb, one layer at a time. Smooth out any wrinkles (Figures 7–20 and 7–21).
4. Cut felt pads to cover bony prominences (especially the patella, fibular head, tibial crest, and condyles if prominent). Secure in place with tape (Figure 7–22).
5. Cut a thin oblong pad, approximately 1-inch wide, to cover the distal posterior calf and proximal posterior thigh. Secure with tape (Figures 7–23 and 7–24).
6. Position the knee at the desired angle of extension, at submaximal range.
7. Begin wrapping with the webril. Start at the knee with a figure-eight around the joint. Proceed distally then proximally, overlapping webril by one-half its width. All areas should have two to three layers of webril. Tear webril to eliminate wrinkles.
8. Grasp the end of the plaster/fiberglass roll. Dip in warm water. Remove and squeeze out excess.
9. Begin wrapping at the knee to set the joint (Figures 7–25 and 7–26).
10. Once the joint is set, proceed with wrapping from a distal to a proximal direction. Leave 1 inch of webril exposed at both ends (Figures 7–27 and 7–28).
11. Smooth plaster with the palms of your hands to avoid indentations (Figure 7–29). (Note that fiberglass does not need to be smoothed.)
12. Repeat procedure with remaining plaster rolls. A total of three layers should be used.
13. Fold back stockinette at both ends and secure with a roll of plaster/fiberglass (Figures 7–30, 7–31, 7–32, and 7–33).
14. Plaster will set in 15 to 20 minutes but will not completely harden for 24 hours. Fiberglass will set/harden in 15 minutes.
15. Assess skin color, temperature, and sensation.
16. Cast should remain in place for 5 to 7 days.

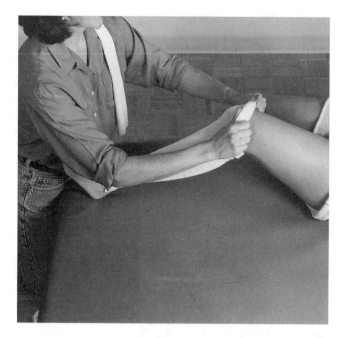

Figure 7–20 Apply stockinette one layer at a time.

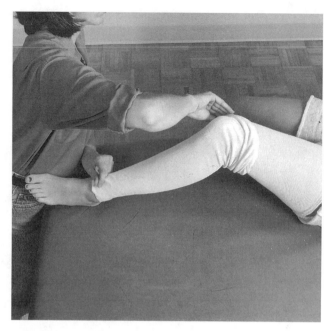

Figure 7–21 Smooth out wrinkles.

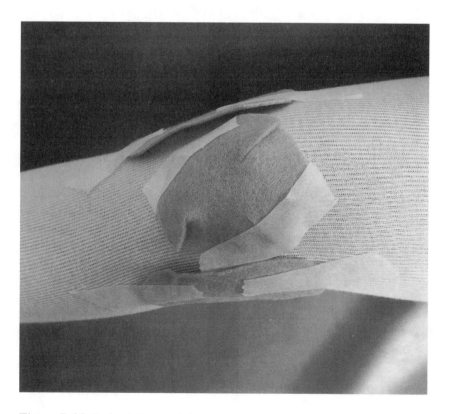

Figure 7–22 Pad patellar condyle.

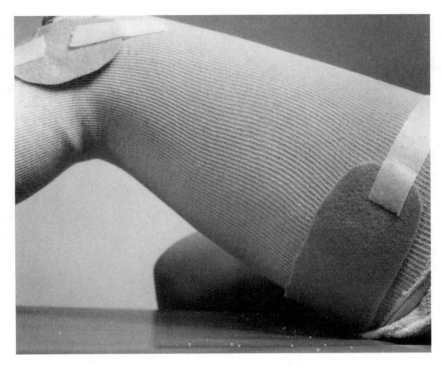

Figure 7–23 Pad posterior thigh.

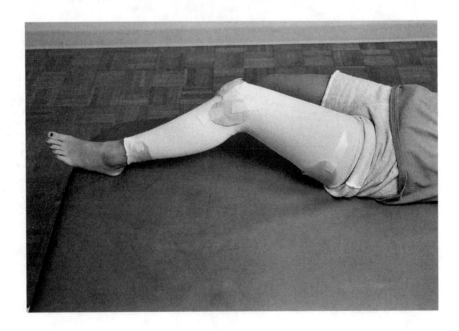

Figure 7–24 Pad distal calf.

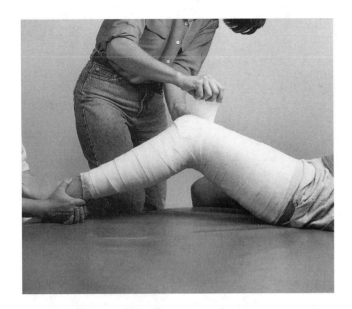

Figure 7–25 Begin wrapping plaster at knee joint.

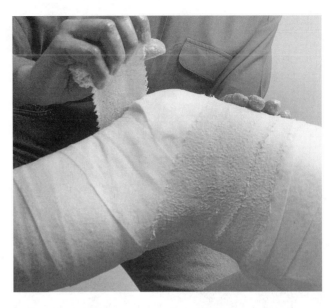

Figure 7–26 Wrap plaster at knee to set joint.

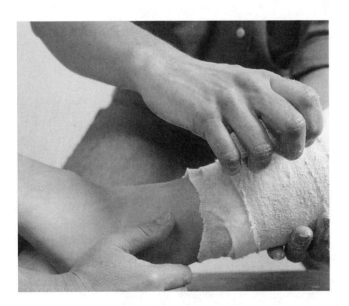

Figure 7–27 Continue wrapping distally.

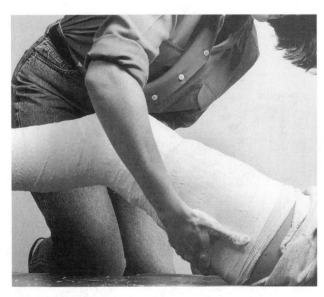

Figure 7–28 Wrap proximally.

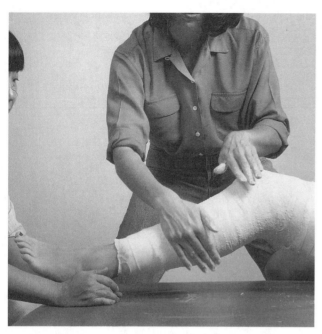

Figure 7–29 Smooth plaster with palms.

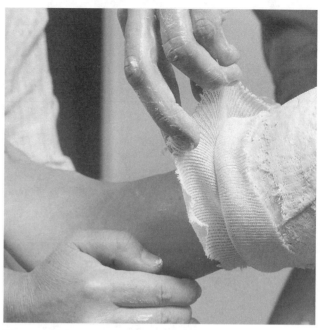

Figure 7–30 Fold back stockinette.

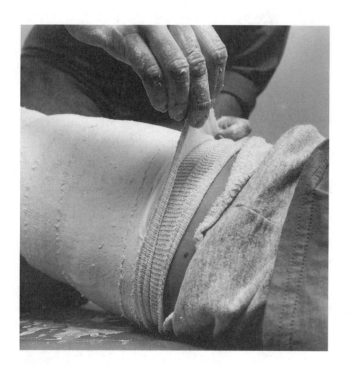

Figure 7–31 Continue folding back stockinette.

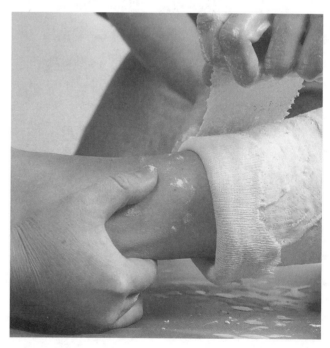

Figure 7–32 Secure ends with plaster.

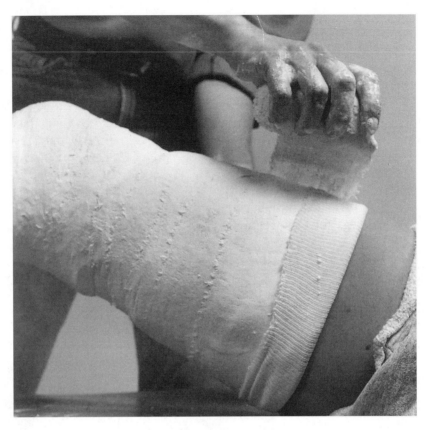

Figure 7–33 Secure ends with plaster.

TONE-REDUCING FOOTPLATE

Rationale

- Allows for control of forefoot, midfoot, and rearfoot.
- Therapist has the ability to build in relief for hypersensitive areas.
- Therapist has the ability to better control the toes.
- Therapist has the ability to control the foot position during the casting procedure.

Prepare Materials

- One piece of cardboard
- Marker
- One box of plaster splints
- Warm water
- Zonas tape

Fabrication Procedure

1. Place the patient in sitting position with hips and knees at 90° and forefoot in subtalar joint neutral.
2. Place cardboard under the foot.
3. Using a marker perpendicular to the cardboard, trace an outline of the foot (Figure 7–34).
4. Mark a dot in the web space between each toe (Figure 7–35).
5. Mark the proximal and distal points of the first and fifth metatarsal fat pads (Figures 7–36 and 7–37).
6. Trace under longitudinal arch with the marker.
7. Mark the peroneal arch, one mark just proximal to the base of the fifth metatarsal and one mark just distal to the fat pad of the heel (Figure 7–38).
8. Remove foot from cardboard.
9. Beginning at the top mark of the first metatarsal (Figure 7–39), draw a line connecting the web

space marks, ending at the fifth metatarsal. These dots define the distal edge of the metatarsal heads.

10. Beginning at the bottom line of the first metatarsal head, draw a line using the marker as if following under the metatarsal heads (Figure 7–40), arching up at the third and dropping down for the fifth. This line marks the proximal borders of contact between the metatarsal heads and the ground (Figure 7–41).

11. Draw an oval at the heel to allow the calcaneous to be seated (Figure 7–42). The oval should end slightly anterior to markings for peroneal arch and navicular.

12. Place patient's foot back on the footplate and check for correct alignment.

13. Cut footplate, following border of the foot.

14. Trace a second footplate; this will serve as the base.

15. Take the original footplate and cut out an oval for the calcaneous and along lines marking the proximal and distal boundaries of the metatarsals (Figure 7–43).

16. Glue these two pieces onto the base (Figure 7–44).

17. Place the patient's foot back on the footplate and check for correct alignment (Figure 7–45). The metatarsal heads and calcaneous should drop down into the depressed areas.

18. Dip small strips of plaster that are 4 to 5 inches in length in water, squeeze out excess, and push under medial longitudinal arch. Continue until arch is filled in, smoothing with fingers as you proceed. Greater volume should be closer to the talus to resemble the arch of the foot. The metatarsal heads should remain seated in the recessed area. Excess amounts of plaster will force the foot to supinate (Figures 7–46 and 7–47).

19. Using the same process, continue with peroneal arch. This arch will be much smaller than the medial longitudinal arch. Excess plaster here will force the foot to pronate. Do not pack under the styloid process because it is a weight-bearing prominence (Figures 7–48 and 7–49).

20. To form the heel cup, take a strip of plaster and measure from the medial arch, along the back of the heel, ending at the peroneal arch. Dip in water, squeeze out excess, and begin at the medial arch. Shape the plaster strip along the border of the heel, connecting at the peroneal arch (Figure 7–50).

21. Place the foot back on the footplate, watching alignment. Check toe position; if flexion is noted, raise toes two through five with plaster to neutral joint position (Figure 7–51). Do not hyperextend the toes.

22. Optional Step: If excessive toe clawing is noted, raise the foot off the footplate and assess the height of the transverse arch (that is, place thumb proximal to second and third metatarsal heads) (Figure 7–52), take a small square of plaster and dip in water, squeeze into a ball and press into depression (Figure 7–53). Now place foot back on footplate, maintaining proper alignment (Figure 7–54). Hold in place for approximately 1 minute so that the plaster binds with the footplate. Carefully lift foot off plate. The finished product looks like Figure 7–55.

23. Smooth any rough edges with a utility knife.

24. Take two long strips of wet plaster and cover the entire footplate (Figure 7–56). Smooth plaster into all recesses and raised area (Figure 7–57). This ties all areas together. Trim away excess plaster.

25. Allow footplate to dry 24 hours prior to weight bearing.

26. Cover footplate with a thin piece of felt.

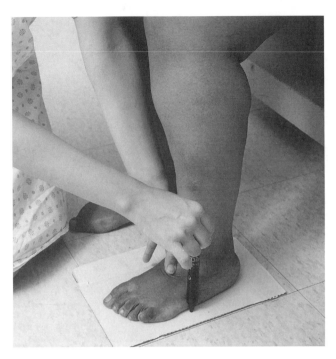

Figure 7–34 Trace outline of foot.

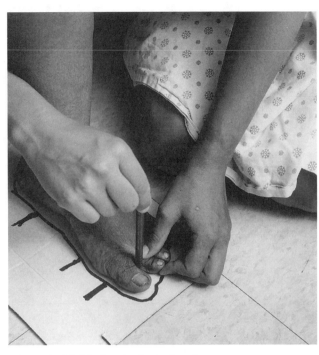

Figure 7–35 Mark a dot in each web space.

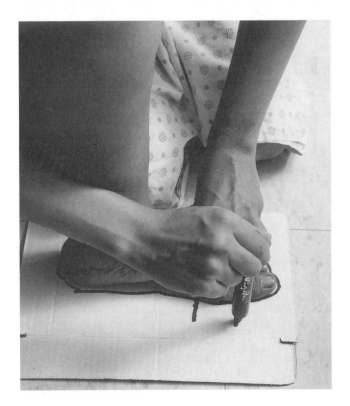

Figure 7–36 Mark proximal and distal points of the first metatarsal.

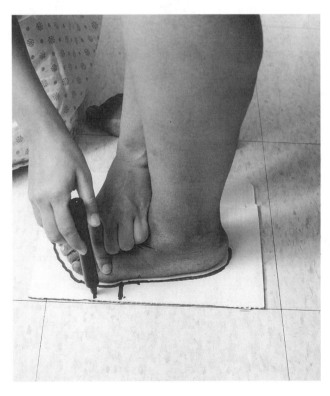

Figure 7–37 Mark proximal and distal points of the fifth metatarsal.

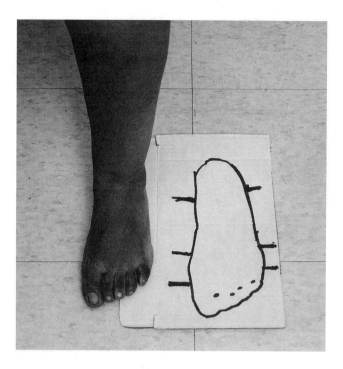

Figure 7–38 Be sure to mark distal medial and lateral first and fifth metatarsal and toe calcaneus.

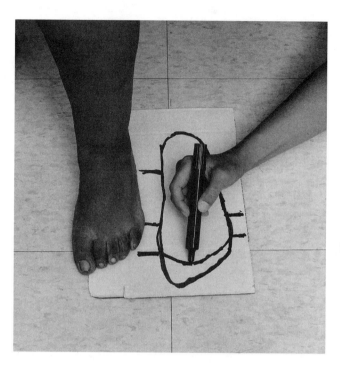

Figure 7–39 Connect the web space dots. The line should arch up gradually and drop down to the fifth head.

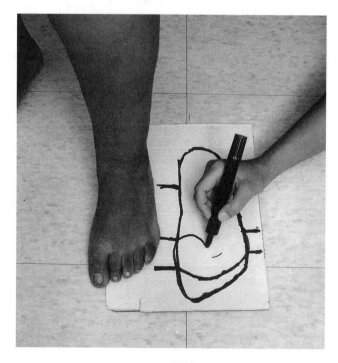

Figure 7–40 Draw a line starting at proximal first metatarsal arching up at third metatarsal.

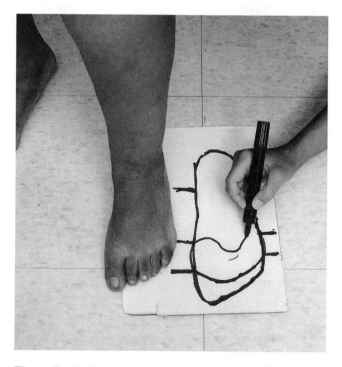

Figure 7–41 Drop line down to proximal edge of fifth metatarsal.

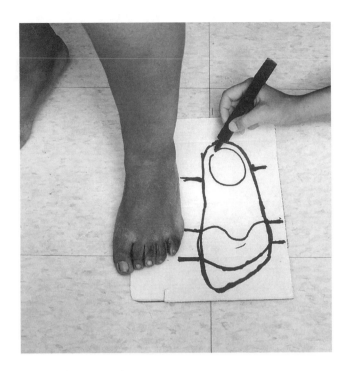

Figure 7–42 Draw an oval to seat calcaneus.

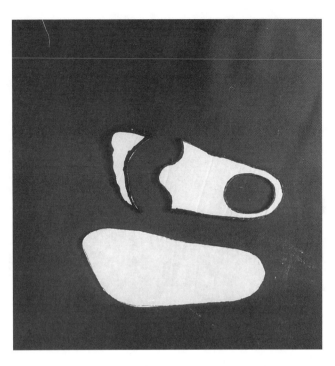

Figure 7–43 One solid foot plate, one recessed footplate.

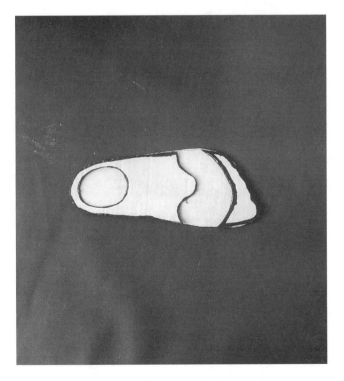

Figure 7–44 Glue two pieces together.

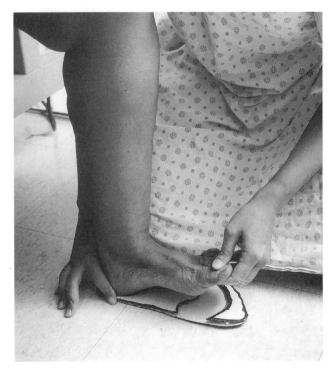

Figure 7–45 Check alignment.

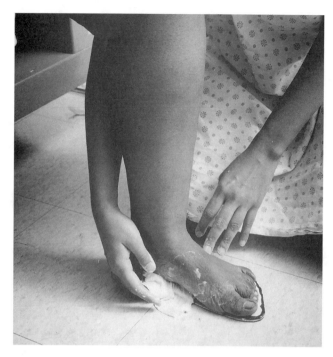

Figure 7–46 Begin filling in medial longitudinal arch.

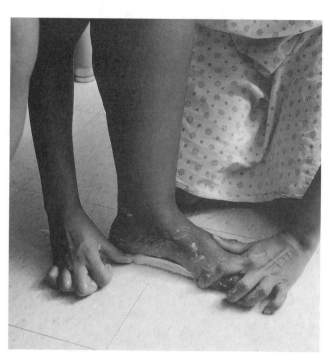

Figure 7–47 Continue until arch is filled. Be sure metatarsal heads remain seated.

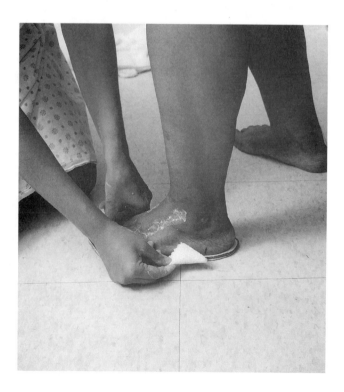

Figure 7–48 Fill in peroneal arch.

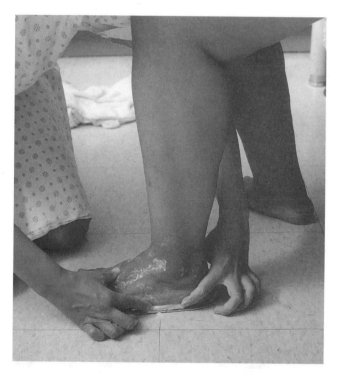

Figure 7–49 Smooth plaster; be sure not to overfill.

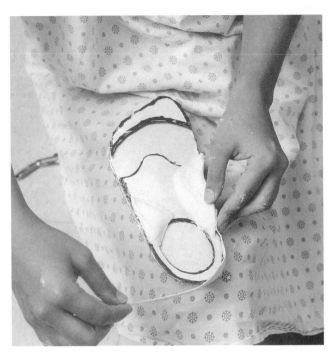

Figure 7–50 Form heel cup.

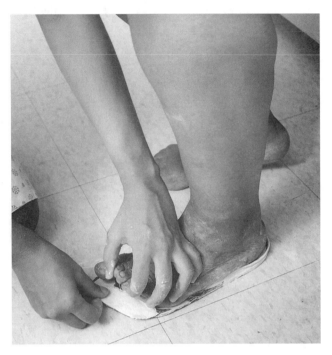

Figure 7–51 Check toe position. If flexed, apply plaster under PIP joints two through five until toes are in neutral position. Do not overextend toes.

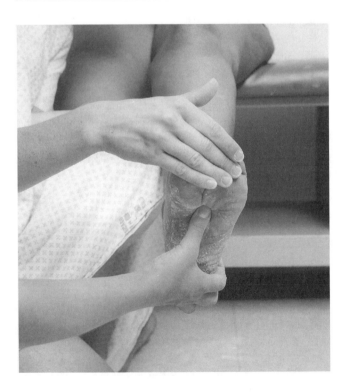

Figure 7–52 Locate transverse arch by applying pressure just proximal to third metatarsal head.

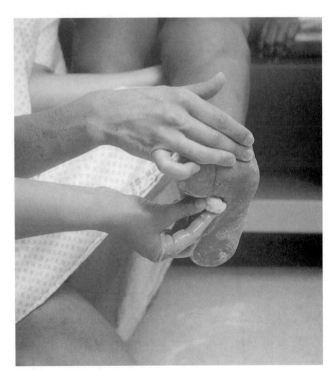

Figure 7–53 Squeeze plaster into a small ball.

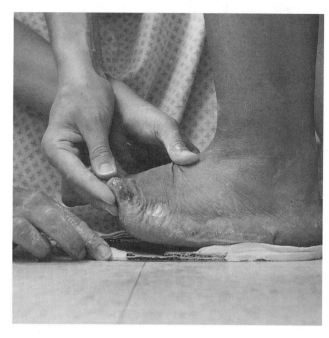

Figure 7–54 Place in transverse arch and carefully place foot back on footplate.

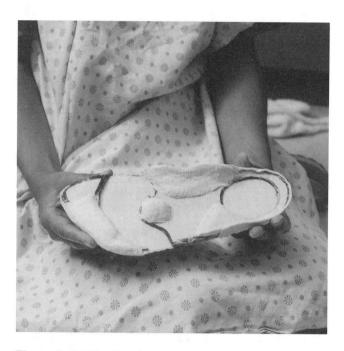

Figure 7–55 The footplate.

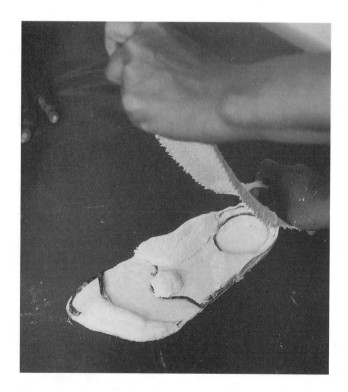

Figure 7–56 Cover footplate with two strips of wet plaster.

Figure 7–57 Smooth plaster into recessed areas.

SUPRAMALLEOLAR ORTHOSIS (SMO)

Rationale

The supramalleolar orthosis (SMO) is used to permit increased freedom of movement at the ankle during gait, while providing medial/lateral stabilization.

Prepare Materials

- Fabricated footplate
- Plaster: 2-, 3-, or 4-inch, three rolls
- Fiberglass: 3-inch, one roll
- Cotton webril: 2-, 3-, or 4-inch depending on size of extremity, three to four rolls
- Felt pads to cover bony prominences
- Stockinette: 2- or 3-inch depending on size of lower extremity
- Tape
- Water
- Scissors
- Cast saw
- Cast cutters
- Cast spreaders
- Towels and gown to cover patient

Fabrication Procedure

1. Follow steps 1 through 4 in Short Leg Serial Cast—Variation I.
2. Apply one layer of felt over the footplate for cushioning (Figure 7–58).
3. Place padded footplate on patient's foot, watching alignment. Secure in place with tape (Figure 7–59).
4. Proceed with steps 5 through 7 in Short Leg Serial Cast—Variation I (Figure 7–60).
5. Wrap with plaster to approximately 3 inches above the malleoli. Make sure that you wrap snugly around the forefoot to ensure that the footplate remains in contact with the sole of the foot (Figures 7–61 and 7–62).
6. Fold down stockinette and finish off as you would in Short Leg Serial Cast—Variation I (Figures 7–63 and 7–64).
7. Reinforce cast with footboard with a roll of fiberglass, for durability during weight bearing.
8. Remove cast as outlined in procedures for cast removal (see Chapter 8).
9. Locate the top of the medial malleolus and estimate the most anterior point (Figure 7–65). Draw a line up (approximately one fingerwidth), curving down and around to the most posterior point, encasing the malleolus. Repeat on the lateral side (Figure 7–66).
10. Optional trimlines
    a. To allow for dorsiflexion/rollover in stance, draw an elliptical line beginning at the anterior point of the medial malleolar line, drop down, and curve back up to the anterior point of the lateral malleolar line.
    b. To allow for free plantarflexion, draw a line beginning at the posterior point of the medial malleolar line, curve down to approximately one fingerwidth above the calcaneous, and back up to the posterior point of the lateral malleolar line (Figure 7–67). (This option should not be used with patients who display a plantar thrust or hyperextend the knee.)
11. Cut along the trimlines with a cast saw. Remove excess cast.
12. Cut down medial and lateral trimlines with bandage scissors, cutting through stockinette, webril, and padding.
13. Finish off cast as outlined in bivalve instructions steps 3 through 10 (see Chapter 8).

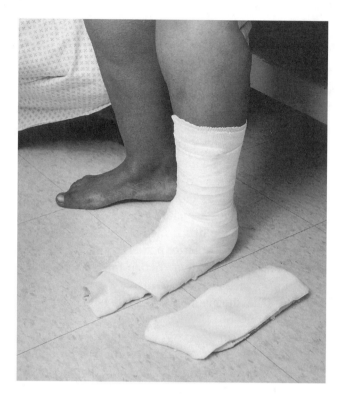

Figure 7–58 Apply one layer of felt over footplate.

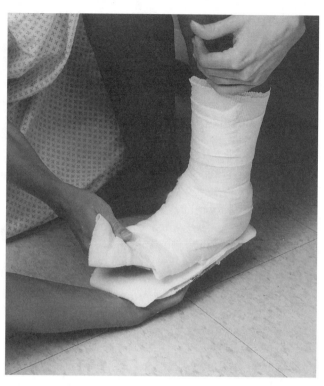

Figure 7–59 Place foot on footplate. Be sure to check alignment.

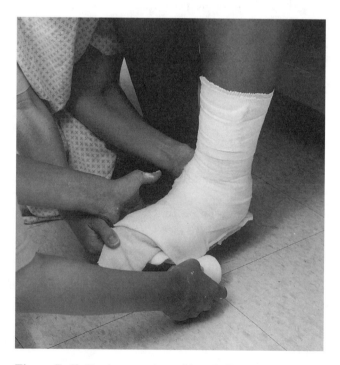

Figure 7–60 Begin wrapping with webril.

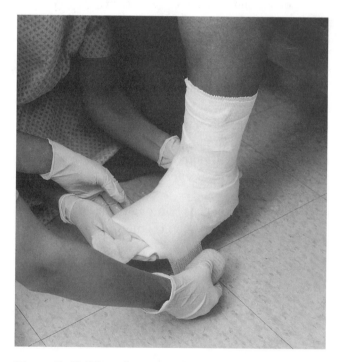

Figure 7–61 Wrap plaster snugly around forefoot.

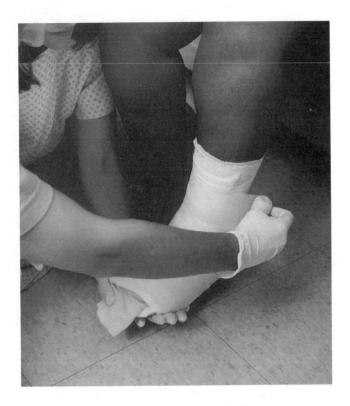

Figure 7–62 Continue wrapping plaster proximally.

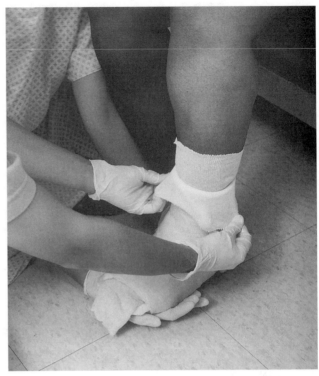

Figure 7–63 Fold back stockinette.

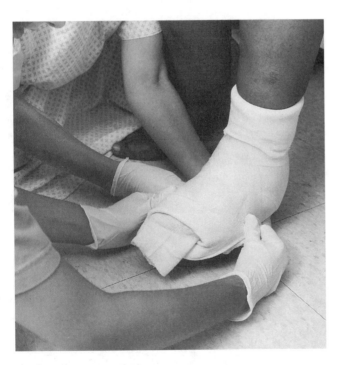

Figure 7–64 Continue folding back stockinette.

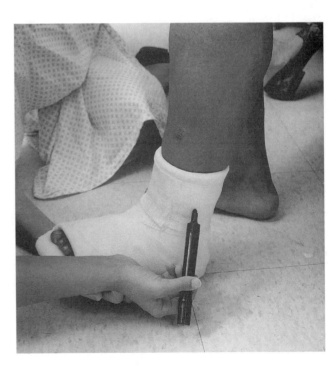

Figure 7–65 Locate top of medial malleolus.

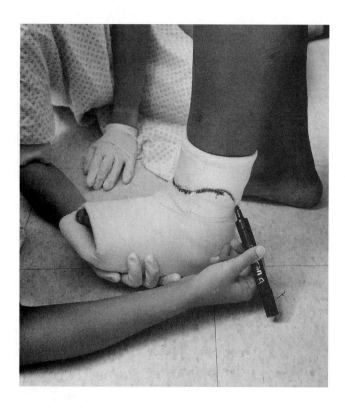

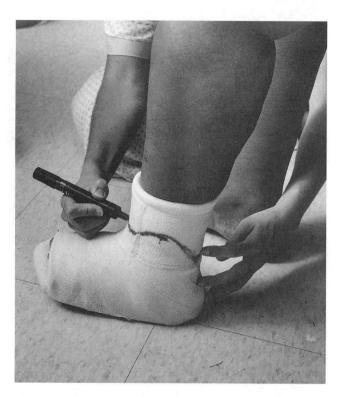

Figure 7–66 Repeat on lateral side. Begin drawing an elliptical line between malleoli.

Figure 7–67 The line should drop anteriorly and posteriorly one fingerwidth below malleoli, which allows for dorsiflexion and plantarflexion.

CHAPTER 8

Cast Removal and Bivalving

Paula Goga-Eppenstein, Judy P. Hill, Terry Murphy Seifert, and Audrey M. Yasukawa

CAST REMOVAL

Therapists involved with cast removal must know how to use the related equipment safely and properly and must know cast-removal techniques. To ensure that staff members are competent to perform cast removal, the department should establish guidelines and a staff training program. In addition, a system of quality control related to the use of the cast saws and their maintenance must be instituted. This will result in equipment that is in good working condition and will help prevent unexpected complications. Patient safety is a priority and should never be jeopardized by poorly functioning equipment. (See Appendix J for a list of vendors of cast-removal equipment.)

Types of Cast Saws

The two types of cast saws most commonly used are the electrically motorized cast cutter (Figure 8–1) and the battery-operated, portable cast cutter (Figure 8–2). To ensure efficiency in cutting and patient safety, use sharp blades only. Also check to be certain that the blade is installed snugly to the cast saw. The blade does not rotate. Instead, it oscillates backward and forward for an excursion of about ⅛ inch. The blade is designed to cut through the plaster or fiberglass and is not meant to cut through padding or stockinette. The cast saw should not cause surface tissue damage or burns when properly used.

Therapists should become familiar with the operating instructions and maintenance for the cast saws that they use. For example, the operating instructions for an electric versus a battery-operated cast cutter are different in the best ways to grip and angle the instruments for optimal control. The motion used to cut the cast is also different. When the full-circle blade is used, an in-and-out motion to move along the line of cut is recommended (Figure 8–3). With the portable, quarter-circle blade, after an initial rocking motion to start the cut, the blade is pulled toward the therapist's thumb at an angle (Figure 8–4). If using both types of cast saws, dedicate one for use with fiberglass and one for use with plaster. At our facility, the electric cast cutter with the vacuum attachment is used primarily for cutting fiberglass casts, and the blade is changed more often due to dulling of the blade. The battery-operated cast saw is used for removing plaster casts. There are also two different types of blades. One blade is designed for use with plaster casts, and its teeth are in line. The blade designed for use with fiberglass casts has offset, serrated teeth. Another helpful tip: We have noticed that children are less fearful of the battery-operated cast cutter because it is much quieter and the saw and blade are smaller.

Equipment and Materials Needed for Cast Removal

- Cast cutters
- Marker

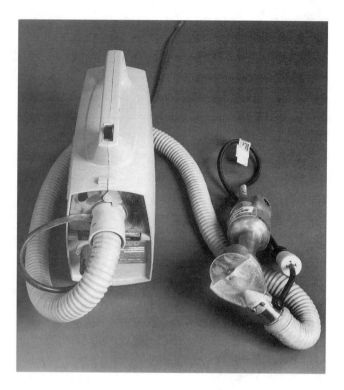

Figure 8–1 Electrically motorized cast cutter.

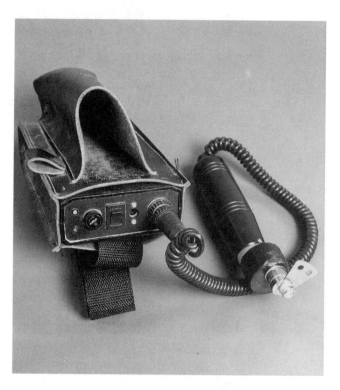

Figure 8–2 Battery-operated, portable cast cutter.

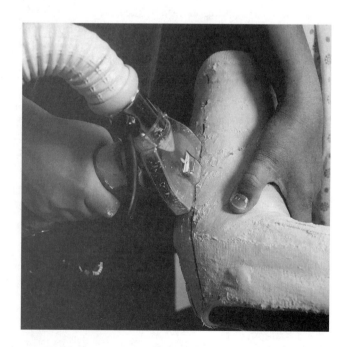

Figure 8–3 In- and out-motion is used with circular blade.

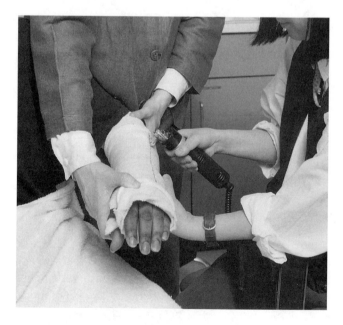

Figure 8–4 Battery-operated cast saw blade is pulled toward the therapist at a 45° angle to cut.

- Cast spreader
- Bandage scissors
- Towel or sheet to drape the patient
- Safety glasses for the therapist and patient—especially important when cutting through fiberglass
- A vacuum attachment on the cast saw or a face mask for filtering potentially harmful airborne dust associated with plaster or synthetic material

Technique

Keep these tips in mind prior to removing a cast. Another person may be needed to assist with holding the casted limb during the cast's removal. Position the patient comfortably either in sitting position or supine position on the plinth. Determine your trimlines. Draw lines along the medial and lateral aspects of the cast, dividing the cast in halves into an anterior and a posterior shell (Figure 8–5). If the cast is to be used as a bivalve, see next section on bivalving casts to determine the proper trimlines.

The loud sound of the cast saw may be extremely frightening to a patient. Before using the cast cutter, the therapist should describe the cast-removal procedure and reassure the patient that the cast blade will not cut into the skin. The therapist can turn the saw on to let the patient hear and see it. Also, the blade can actually be touched lightly to the skin to demonstrate that it does not cut (Figure 8–6).

Use of Cast Cutters

Electric Motor Cast Cutter

1. Hold the electrically powered cast cutter, using a comfortable grip on the narrow part of the cutter next to the blade (Figure 8–7).
2. Hold the limb still during cast removal. Starting at the top or bottom of the cast, place your fingers inside the cast to move soft tissue away from the cast. Hold the cast saw with the blade perpendicular to the cast (Figure 8–8). Press down firmly until you feel the material give or you feel an ease of resistance. This indicates that you have cut through the hard plaster or resin material.

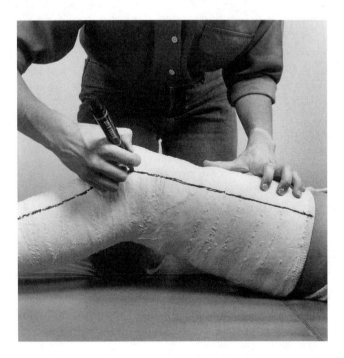

Figure 8–5 Draw trimlines to divide cast into anterior and posterior shells.

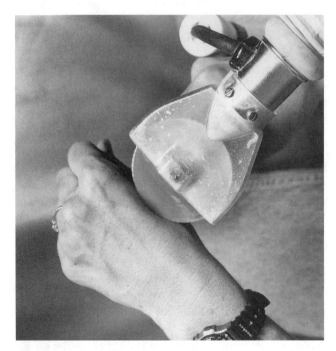

Figure 8–6 Therapist demonstrates to patient that blade touched lightly to the skin does not cut.

Figure 8–7 Use a comfortable grip on the narrow part of the cutter.

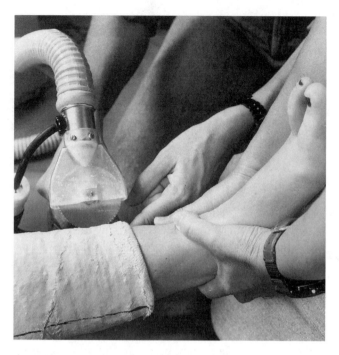

Figure 8–8 Hold cast saw with blade perpendicular to cast.

3. Pull the cast saw up slightly and continue along the trimline.
4. Move the blade forward and repeat, cutting vertically up and down. DO NOT pull the blade parallel to the line as this may result in tearing through the padding and searing the patient's skin. DO NOT hold the blade in the same position for extended periods because this could burn the patient. Check the cast saw's temperature. If it is hot, turn it off and allow the motor to cool.

Battery-Operated Cast Cutter

1. Hold the battery-operated cast cutter, using a comfortable grip (Figure 8–9).
2. Cut through the cast, initially applying pressure downward, perpendicular to the blade and rocking the blade slightly.
3. Cut through the plaster or fiberglass and lift the blade up slightly.
4. Angle the blade at 45° and pull it along the line of cut.
5. Let the saw do the work; never force it.

With this cast saw, since the blade is held at an angle, the blade teeth are not oriented toward the skin. Therefore, pulling the blade along the line of cut does not pose the same risks as with the circular blade. *Always consult manufacturer's instructions before using a cast saw.* It can be helpful for the beginner to practice technique on a cast that has already been removed from the limb to gain confidence in using the saw.

Insert the spreader between the sawed edges of the cast and spread the edges apart (Figure 8–10). If resistance is felt and it will not separate, locate problem spots and recut with the cast saw. Use the bandage scissors to cut the stockinette that was pulled up over the distal and proximal ends of the cast. Pull the anterior shell off. Rip or cut through the remaining cast padding, allowing the scissors to glide along the stockinette (Figure 8–11). Use the bandage scissors to cut through stockinette (Figure 8–12). Remove the padding. Wash the limb and follow the protocol for assessment.

BIVALVING CASTS FOR USE AS SPLINTS

A bivalved cast can serve as a maintenance orthosis to preserve the gains made by the serial casting tech-

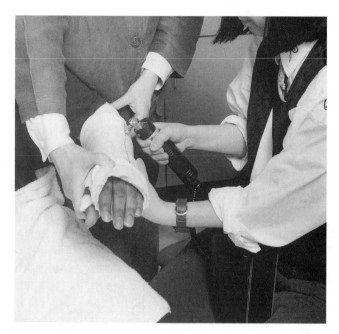

Figure 8–9 Use a comfortable grip.

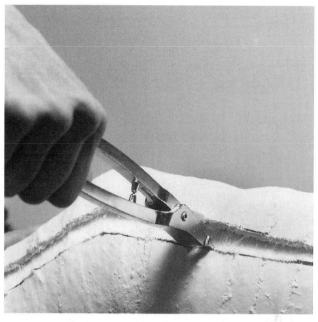

Figure 8–10 Insert spreader between sawed edges of cast.

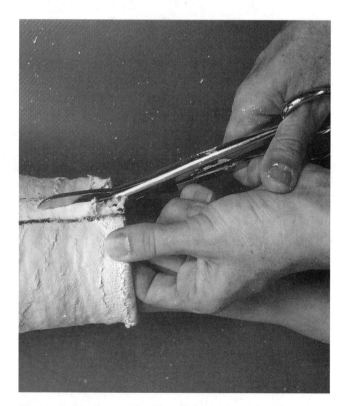

Figure 8–11 Cut padding by gliding scissors on the stockinette.

Figure 8–12 Cut through the stockinette.

nique. The purposes of using a bivalve cast may include the following:

- to maintain passive range of motion and prevent further joint deformity
- to permit active motion and participation in a program of exercise
- to access the limb for hygiene
- a limited ability to monitor tolerance of the cast, as occurs with outpatients

Bivalved casts are generally used as positioning devices intermittently throughout the day and night. Ideally they are worn at night only, so that the patient can participate in activities during the day. Occasionally, however, they are used during periods of rest during the day as well. Long-term use for more than 12 hours per 24-hour period is generally not recommended, as this will contribute to disuse atrophy of muscles and result in considerable joint stiffness. An individual who continues to require the cast to be in place 12 to 24 hours per day to maintain range must be considered for alternate management, including nerve block and surgery.

Bivalve casts may be used in place of serial casts when special considerations exist such as when the patient presents with poor sensation and skin breakdown is a concern. If a bivalve cast is used, it will allow for frequent skin checks. When a patient has poor tolerance of a serial cast, bivalving will allow the patient to have time off from the restrictions of the cast. When a limited ability to monitor tolerance of the cast exists, as occurs with outpatients, using the bivalve cast as a serial cast will allow for quick removal if problems arise. When the bivalved cast is being used as a serial cast, the plan is to leave it in place for the full 5 to 7 days. The bivalve is removed only if the patient is not tolerating the cast or an emergency arises. The bivalve cast must fit snugly and comfortably on the patient's limb to ensure its optimum effectiveness. If the bivalve is poorly tolerated by the patient and pressure areas are noted after removal of the cast, the clinician must reevaluate the fit. Several problems may occur during fabrication of the bivalve cast. For instance, the cast may have been applied with the joint and muscles in maximal stretch. Positioning in maximal range does not allow the patient gradually to build up wearing tolerance. The bivalve cast must be fabricated at

submaximal range or at a position slightly less than maximal range of the joint to ensure comfort. If it is not, recurrences of the contracture are more likely to occur because of the poor compliance associated with improper fit. Ideally a bivalve is made from a cast that has been left on the limb from 2 to 5 days, allowing soft tissues time to accommodate to the position.

Once a cast is bivalved, the original integrity is lessened. This may be due to the fit of the finished product or poor strap adjustment. Be aware of who will be applying and removing the bivalve cast each day. Be sure to advise him or her about correct application and strap placement. The potential for skin breakdown often increases when the cast is bivalved. If the splint is not applied correctly, there is a risk of compressing skin between the anterior and posterior parts of the splint and increased pressure points due to poor alignment of the splint on the limb. Prior to implementing a program for wearing the bivalve cast the therapist must check out the following areas:

- Is the bivalve fabricated at the proper range for a good fit?
- Does the bivalve fit snugly? (Is it too big or too tight?)
- Are the padding and rough edges of the bivalve secured with tape?
- Are the straps securely attached to the bivalve?
- Have the two sides of the cast been clearly marked to match and been correctly aligned?

Once the bivalve cast is checked out for proper fit, the therapist must instruct the patient or caregiver about its proper application. Instruct the caregiver on techniques for relaxing the patient's extremity prior to donning the bivalve cast. Once the extremity is relaxed, place it into the anterior and posterior shell of the bivalve cast, then secure the cast together with the straps. The patient or caregiver must be able to demonstrate how to apply the bivalve correctly to ensure proper fit and wearing tolerance. In addition, he or she must express an understanding of how to use the device outside the clinic setting. If several caregivers will assist with application of the bivalve cast, an instruction sheet as well as labeling of the bivalve cast will be helpful. (Use a permanent marker to label the top and bottom of the cast, or the right and left if both sides are cast.)

Upper Extremity Bivalve Cast Procedure

The bivalve cast can be fabricated with a lightweight, durable material such as fiberglass. The material is a synthetic casting bandage that generally sets within 3 to 5 minutes depending on the specific instructions given for use. Fiberglass application instructions typically include the following:

1. Apply the fiberglass using the standard technique for applying plaster casting over stockinette, felt pieces, and cast padding. To accommodate for the shrinkage, use extra cast padding or extra stockinette (Figure 8–13). This can later be removed in the bivalve fabrication process.
2. Utilize the general principles and techniques for applying the specific cast being fabricated as described for a cast made of plaster of Paris. Fiberglass is a more rigid material and is not as pliable as plaster. As a consequence, the technique for fabricating a fiberglass wrist cast differs slightly from that for a cast formed with plaster. For example, instead of the figure-eight wrap around the thumb, a vertical slit is cut in the fiberglass

material (Figure 8–14). The thumb is then placed through the hole (Figures 8–15 and 8–16).

3. Wear gloves while handling fiberglass, since the resin will adhere to the skin (Figure 8–17).
4. Remove the roll from its sealed package and submerge it in room-temperature water. Follow the instructions for use regarding the set time factor. To have more set time, use the roll directly from the package; do not submerge the fiberglass in water. Once the cast is applied, a wet Ace bandage can be rolled directly on the contoured fiberglass to help the cast set.
5. Wrap with minimal pull and avoid excessive tension on the roll, since fiberglass has a slight tendency to shrink (Figure 8–18).
6. Smooth and rub the surface of the cast to contour and shape according to the standard technique. For example, form a palmar arch (Figure 8–19). Apply pressure on the volar surface with index, middle, and ring fingers and counterpressure at the wrist, using the volar surface of your hand.
7. Allow the fiberglass cast to set thoroughly prior to removal.

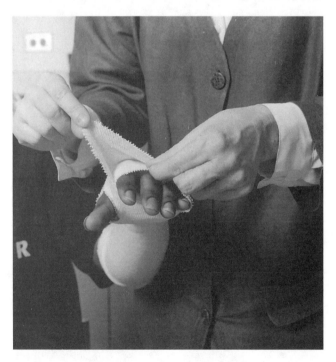

Figure 8–13 Two layers of stockinette can be used to accommodate for shrinkage.

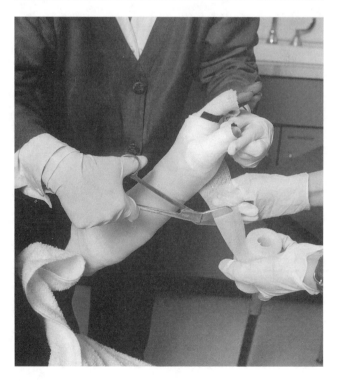

Figure 8–14 Cut a vertical slit in the fiberglass material.

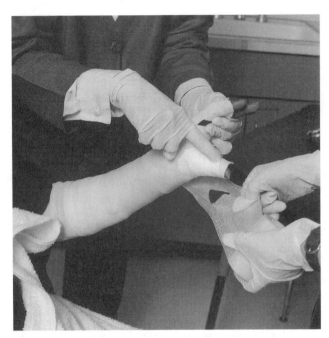

Figure 8–15 Place thumb through hole.

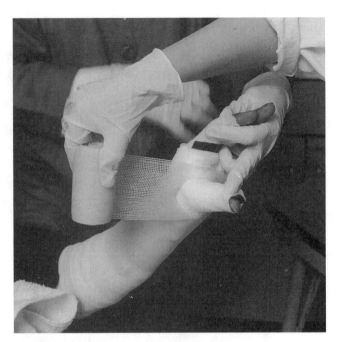

Figure 8–16 Place thumb through hole.

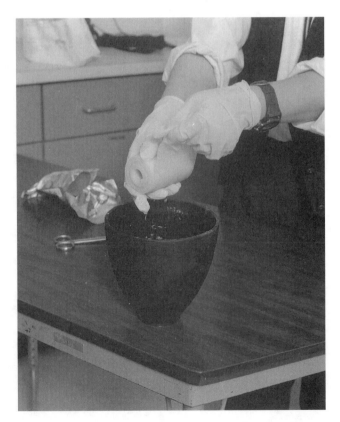

Figure 8–17 Wear gloves while handling fiberglass.

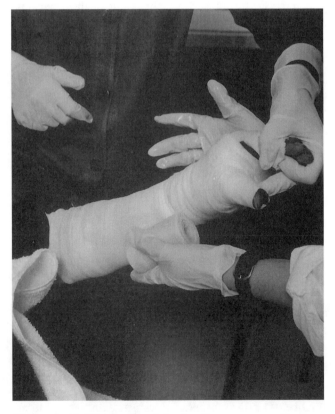

Figure 8–18 Wrap with minimal pull.

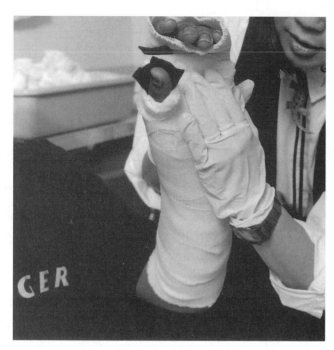

Figure 8–19 Form a palmar arch.

Removal Procedure for the Bivalve Cast

The procedure for bivalving will depend on the type of cast applied. The most commonly used casts are described.

Rigid Circular Elbow Cast

1. Draw two cutlines, one from the axilla to the medial epicondyle, the other from the lateral humerus to the lateral epicondyle, to form the anterior and posterior shell (Figure 8–20). In general, remember that the shells should be left intact against the direction of pull of the limb. For example, when casting the elbow into extension, avoid cut lines and seams over the olecranon and antecubital fossa.
2. Remove the fiberglass cast with a cast saw, following safety precautions and procedures. Safety goggles for the client and therapist are necessary to prevent fiberglass particles from entering the eyes. Using a vacuum attachment with the cast saw is also helpful for filtering the dust particles.

Do not hold the cast blade into the fiberglass for more than a few seconds in any one spot. The blade may become somewhat hot after prolonged use and give a burning sensation. Do not use dull or damaged blades. Be sure to cut on your predrawn cutlines and separate the fiberglass gently with the cast spreader.

3. Let the cast scissors glide on top of the stockinette and cut through the padding. Then cut through the stockinette. Remove one layer of stockinette of the additional layers of padding so that the cast is not too tight. Smooth the rough edges of the fiberglass with a file if needed. To cover the rough edges, tape the padding and edges of the shell (Figure 8–21).
4. The following methods can be used for finishing the bivalve:
   Method A—The original stockinette and padding can be secured to the shell of the Elasticon tape or moleskin.
   Method B—Remove the original stockinette. Cover the anterior and posterior shells with new stockinette. Pull the entire 3-inch or 4-inch stockinette over the anterior and posterior shell (Figure 8–22). Make sure the stockinette is pulled well into the interior of the cast just on top of the padding, not bridging across the shell from edge to edge. Secure the ends of the stockinette with tape. This method provides a smooth covering of stockinette over the rather rough cast.
5. The bivalve shell can be completed and secured together by using D-ring straps attached to the surface of the cast if Method A was used or to the stockinette covering if Method B was used (Figure 8–23).

Another option of applying straps directly onto the synthetic cast with Method A is to apply prefabricated D-ring straps around the completed bivalve and fix them in place using Aquaplast. Heat a small strip of Aquaplast and place vertically over the Velcro strap and then press the Aquaplast into the fiberglass cast. The heated Aquaplast will impregnate into the knitted synthetic material and act as a rivet to hold the strap securely in place.

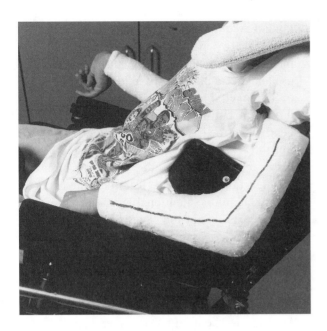

Figure 8–20 Draw two cutlines.

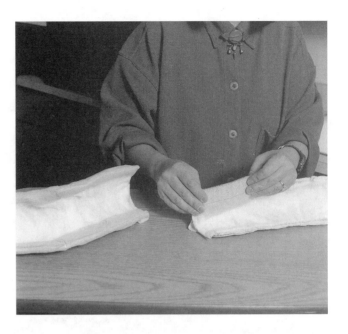

Figure 8–21 Tape the padding and edges of the shell.

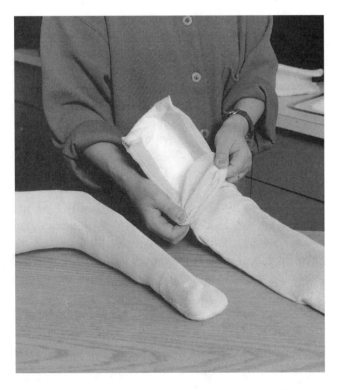

Figure 8–22 Pull the entire stockinette over the shell.

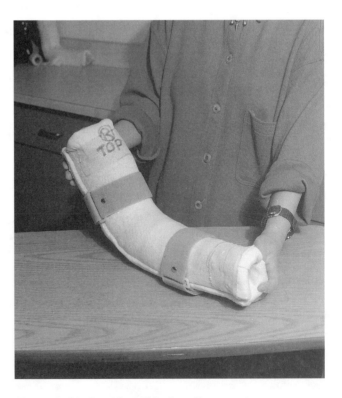

Figure 8–23 Completed bivalve elbow cast.

Wrist Cast

Two options exist for creating anterior and posterior shells when bivalving a wrist cast. Both require making cuts to keep the web space and thumb hole intact. The first offers a more sturdy bivalve with very secure fit. However, of the two, it is the more difficult to get back on. The second is slightly less durable but much easier to place on the limb.

Option 1

1. Draw two lines distally from the fourth metacarpal on the dorsal and volar surfaces of the wrist cast proximally up the forearm (Figures 8–24 and 8–25).
2. Cut through the cast as described above (Figure 8–26). One third of the cast will form a small ulnar gutter and the remaining two thirds of the cast will form a radial gutter. Carefully spread the fiberglass apart with a cast spreader (Figure 8–27). Cut the padding, leaving the stockinette intact (Figure 8–28).
3. Remove the ulnar gutter portion with the padding. Cut through the stockinette on the ulnar portion (Figure 8–29). Carefully remove the radial portion of the cast, leaving the extra stockinette in place (Figure 8–30).
4. The following options may be used for finishing the radial portion of the bivalve.

a. Smooth all rough edges of the fiberglass. Secure padding first and then original stockinette of the radial portion with tape (Figure 8–31).
b. Remove original stockinette. Secure the padding of the radial portion with tape. Apply new stockinette (Figure 8–32). Tug on the stockinette (Figure 8–33) so that it lays smoothly and snugly into the interior of the cast just on top of the padding (Figure 8–34). Cut a small slit in the stockinette for the thumb hole (Figure 8–35) and secure it with tape (Figure 8–36). Secure the ends of the stockinette with tape. (See the completed radial portion of the cast in Figure 8–37.)

5. Secure padding of the ulnar gutter portion with tape (Figure 8–38). Apply new stockinette (Figure 8–39). Secure ends with tape.
6. Apply D-ring straps to secure the wrist bivalve (Figure 8–40).
7. Draw lines for notching the cast with permanent markers to indicate exact alignment of the two halves (Figure 8–41).
8. Cut a small Aquaplast strip. Heat it and press it into the fiberglass portion of the cast to secure the D-ring straps. If stockinette is covering the cast, cut a small portion of the stockinette above and below the straps so the Aquaplast can impregnate the synthetic material and act as a rivet (Figure 8–42).

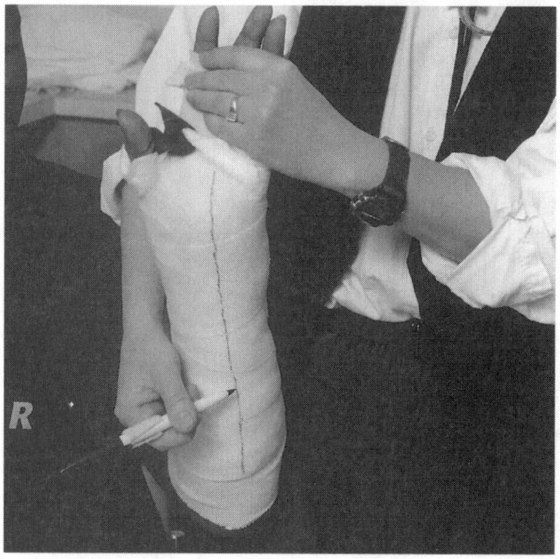

Figure 8–24 Draw line from fourth metacarpal up forearm.

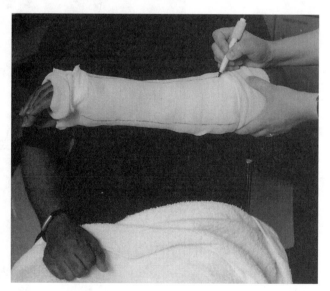

Figure 8–25 Draw line from fourth metacarpal up forearm.

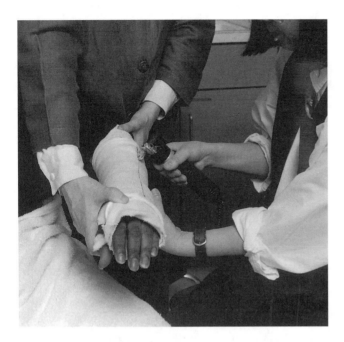

Figure 8–26 Cut through cast.

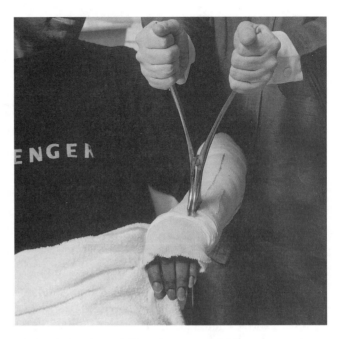

Figure 8–27 Spread fiberglass apart with a cast spreader.

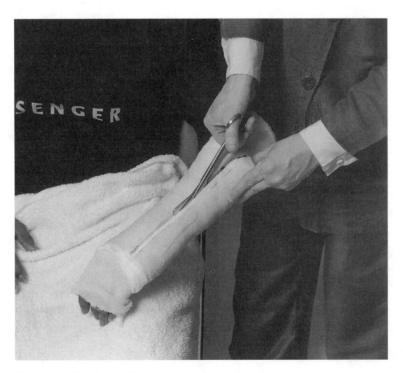

Figure 8–28 Cut padding.

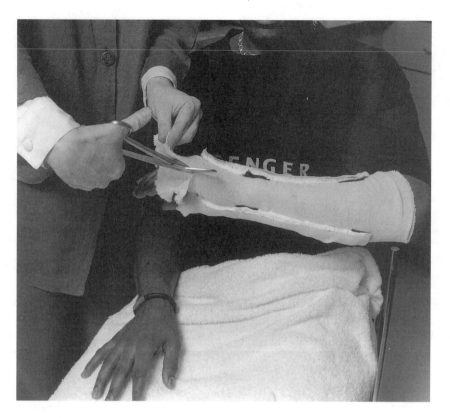

Figure 8–29 Cut the stockinette on ulnar portion.

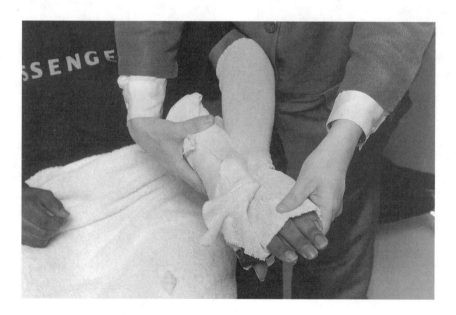

Figure 8–30 Remove radial portion.

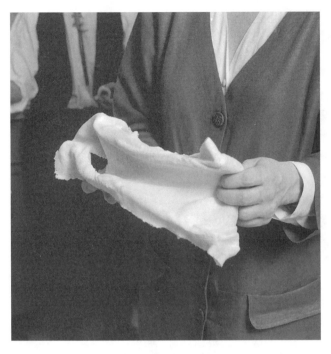

Figure 8–31 Secure padding and stockinette with tape.

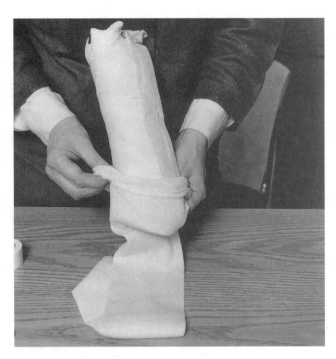

Figure 8–32 Apply new stockinette.

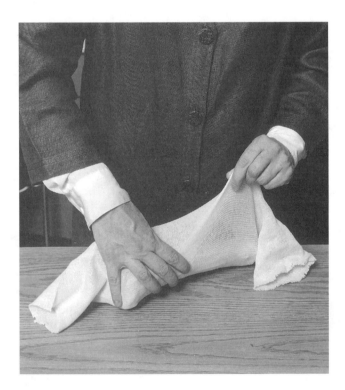

Figure 8–33 Tug on stockinette.

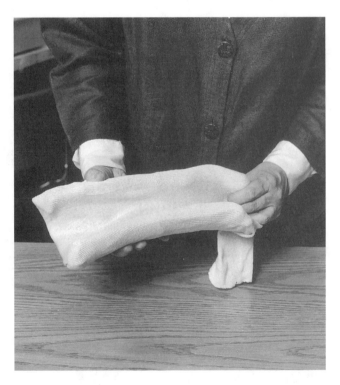

Figure 8–34 Lay smoothly into interior of cast.

Figure 8–35 Cut slit in stockinette for thumb hole.

Figure 8–36 Secure with tape.

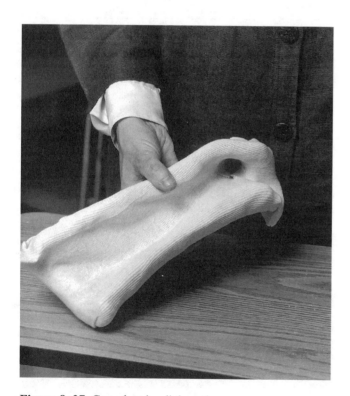

Figure 8–37 Completed radial portion.

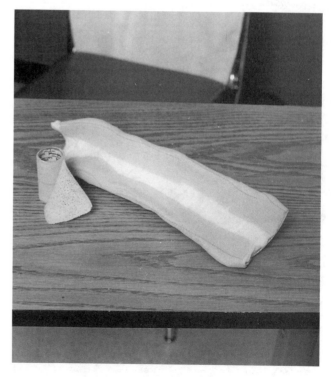

Figure 8–38 Secure padding of ulnar gutter with tape.

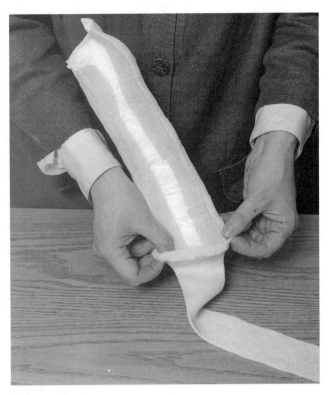

Figure 8–39 Apply new stockinette.

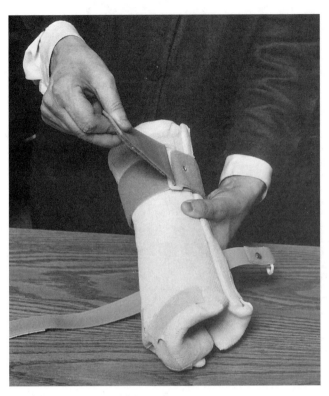

Figure 8–40 Apply D-ring straps.

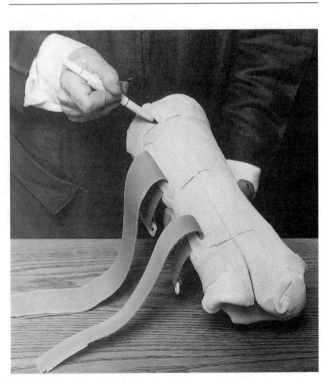

Figure 8–41 Draw lines for notching.

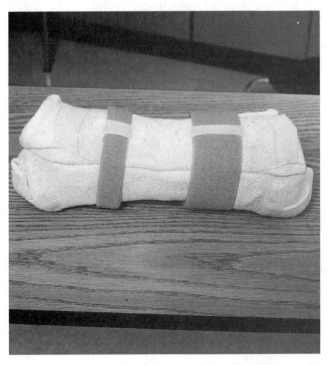

Figure 8–42 Aquaplast impregnated into fiberglass.

Option 2

1. Draw two cutlines. One line follows the lateral fifth metacarpal and ulnar border (Figure 8–43). The other starts at the dorsal second metacarpal, follows the second metacarpal on the dorsal surface to the wrist, then sweeps volar to follow the lateral radial border proximally (Figure 8–44).

2. Cut the stockinette, leaving a ½-inch border on each side of the volar shell. The original stockinette will be left in place on the volar shell. Secure it to the shell with tape. Replace the stockinette on the dorsal shell and secure it with tape.

3. Place D-ring straps at the proximal end of the cast and at the wrist. A third strap at the distal end of the cast through the web space may be necessary.

Long Arm Cast

1. The markings on the long arm cast will be slightly curved to maintain the integrity of the olecranon and web space. Use essentially the same cutlines as the elbow and wrist casts. Mark the medial and lateral epicondyles and the dorsal and volar surfaces in line with the fourth metacarpal. Use the distal and proximal landmarks to draw the lines to form the bivalve (Figures 8–45 and 8–46).

2. Carefully cut the fiberglass, padding, and stockinette. Inspect the edges of the fiberglass. Add or remove any padding and secure the edges as described previously.

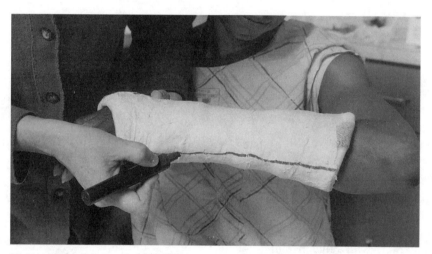

Figure 8–43 Line on lateral fifth metacarpal and ulnar border.

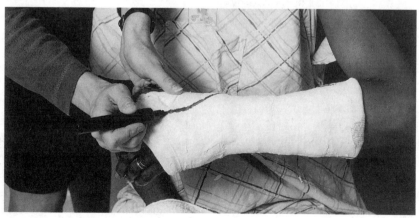

Figure 8–44 Line on radial border.

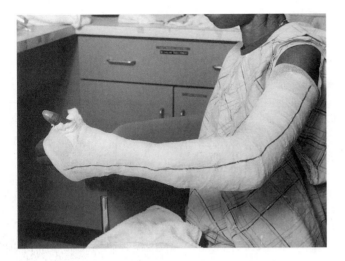

Figure 8–45 Lateral epicondyle to fourth metacarpal.

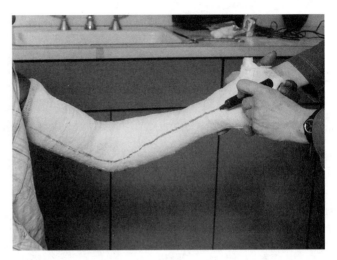

Figure 8–46 Medial epicondyle to fourth metacarpal.

Ankle Bivalve Cast

Materials: Bandage scissors, porous tape, stockinette, Velcro, staple gun.

Fabrication Procedure:

1. See cast removal procedure, items a to e, listed below:
   a. Locate malleoli and mark with a dot (Figure 8–47). Locate fifth toe and mark with a line lateral to the fifth toe, extending the line up to the malleoli (Figures 8–48 and 8–49). Locate the first toe and mark with a line medial to the first toe, extending the line up to the malleoli.
   b. Draw lines along the medial and lateral aspects of the cast up toward the knee, dividing the cast in halves into an anterior and posterior shell (Figures 8–50 and 8–51).
   c. Hold the limb still during removal. Starting at the toe or knee, place your fingers inside the cast to move the soft tissue or toes away from the edge of the cast (Figure 8–52) and cut along trimlines (Figure 8–53).
   d. Insert spreader between the sawed edges of the cast and spread the edges apart (Figure 8–54). If resistance is felt and it will not separate, locate problem spots and recut with the cast saw.
   e. Crimp the edges of the cast to cut the stockinette with the cast cutters (Figures 8–55 and 8–56).

2. To maintain the integrity of padding, use bandage scissors and cut through padding and stockinette, following the trimline (Figure 8–57). Repeat the procedure on the opposite side.
3. Remove anterior and posterior shells (Figure 8–58).
4. Brush out any loose fiberglass.
5. Lay anterior shell on a flat surface. Cut away one layer of stockinette at the top and bottom of the cast, leaving the bottom layer intact.
6. Lay the stockinette on the inside of the shell and stretch it across. Make sure to smooth out any wrinkles.
7. Using porous tape, tape stockinette onto the cast, placing strips lengthwise. Place one or two long strips of tape on the outer shell to secure (Figures 8–59 and 8–60).
8. Do not remove the stockinette from the posterior shell. Smooth out any wrinkles in the existing stockinette. Proceed as in step 7.
9. Place anterior shell on posterior shell.
10. Cut strips of Velcro to place at 45° at the heel, the top of the calf, and across the toes (Figures 8–61 and 8–62). Staple straps in place. Be sure to label the bivalve (i.e., right and left, top and bottom, ankle and calf strap, etc.) to avoid confusion for the caregiver who will be applying it.

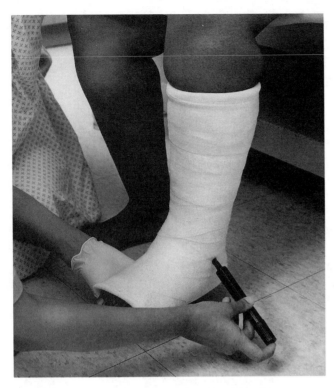

Figure 8–47 Locate malleoli and mark with a dot.

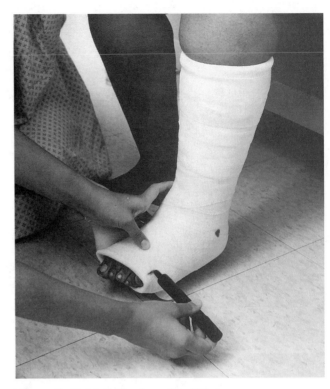

Figure 8–48 Mark line lateral to fifth toe.

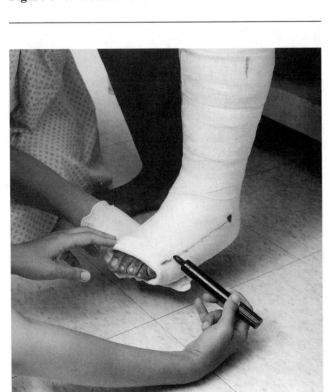

Figure 8–49 Extend line up to the malleoli.

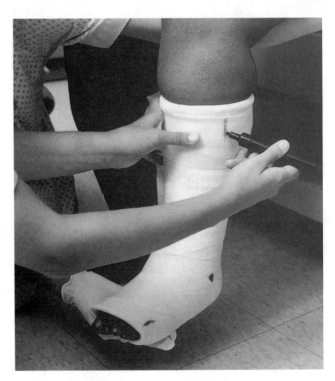

Figure 8–50 Draw line up toward knee.

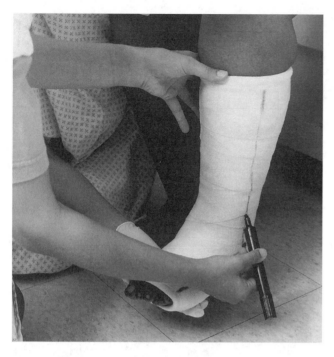

Figure 8–51 Draw line on medial and lateral aspect of cast.

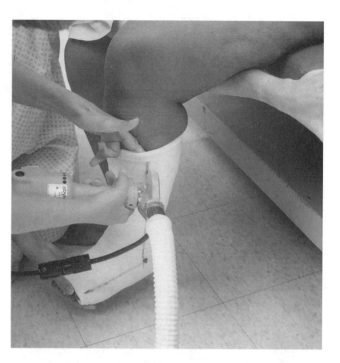

Figure 8–52 Insert finger inside cast to move aside soft tissue. Begin cutting along trimline.

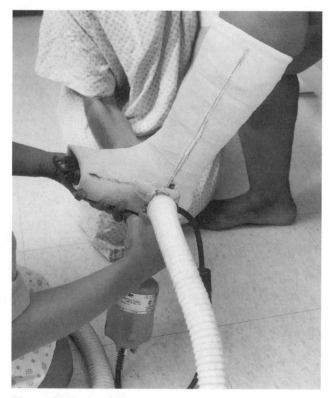

Figure 8–53 Continue cutting along trimline.

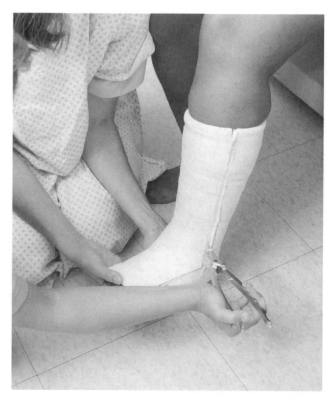

Figure 8–54 Insert cast spreader and spread edges apart.

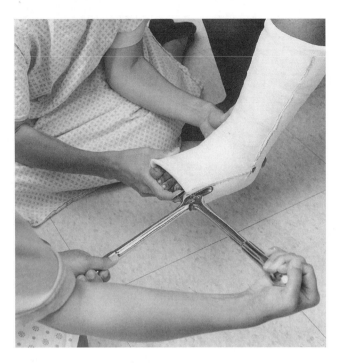

Figure 8–55 Crimp distal edges of cast to cut stockinette.

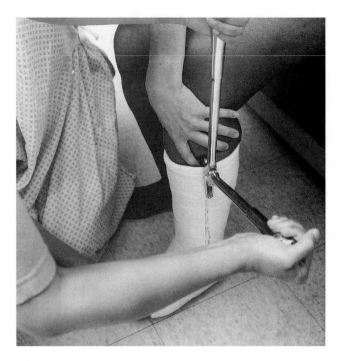

Figure 8–56 Crimp proximal edges of cast to cut stockinette.

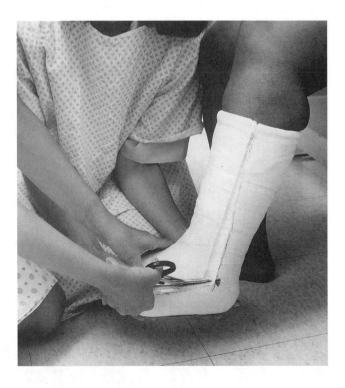

Figure 8–57 Use bandage scissors to cut stockinette and padding. Follow trimline.

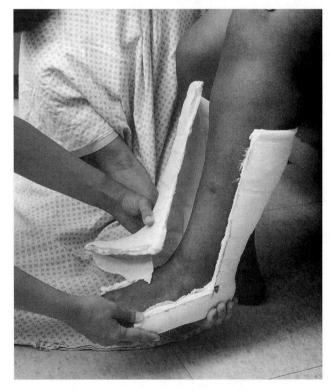

Figure 8–58 Remove anterior and posterior shells.

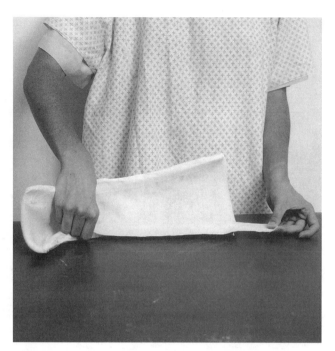

Figure 8–59 Tape stockinette onto cast.

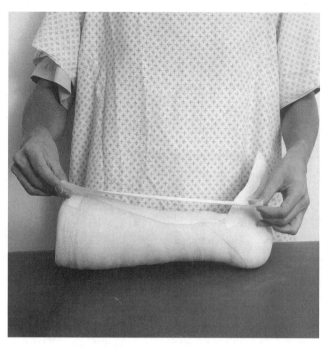

Figure 8–60 Use one or two strips of tape on outer shell to anchor stockinette firmly.

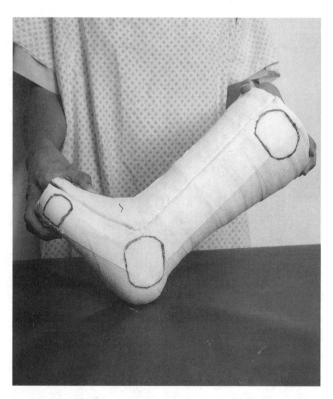

Figure 8–61 Place Velcro at 45° angle to ankle at top of calf and at toes.

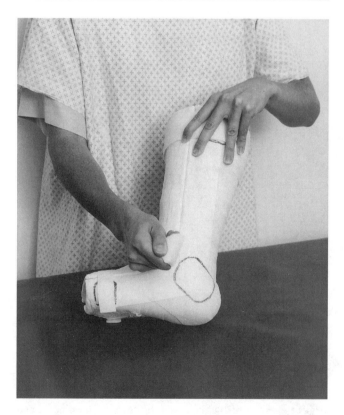

Figure 8–62 Staple straps into place.

Knee Bivalve Cast

Materials: bandage scissors, porous tape, stockinette, Velcro, staple gun.

Fabrication Procedure:

1. See step 1 of Ankle Bivalve Cast fabrication procedure and follow the procedure for cast removal, items a through e, except for the following differences:

   • Locate the medial and lateral condyle and mark with a dot on each side. At the ankle, mark with a line extending up to the condyle. Repeat on the opposite side of the cast (Figure 8–63).

   • At the upper thigh, mark with a line extending down to the condyle. Repeat on the opposite side (Figure 8–64). The bottom, or posterior shell, should be slightly deeper (approximately two thirds the size of the anterior shell) for ease in donning the splint when used as a bivalve cast (Figure 8–65).

2. Follow steps 2 through 9 of the Ankle Bivalve Cast fabrication procedure (Figures 8–66, 8–67, 8–68, 8–69, 8–70, 8–71, 8–72, 8–73, 8–74, 8–75, and 8–76).

3. Cut Velcro strips to fit at the proximal and distal ends of the cast and one strip just proximal or distal to the knee, depending on the degree of flexion. Be sure to label the shell and all straps to avoid any confusion.

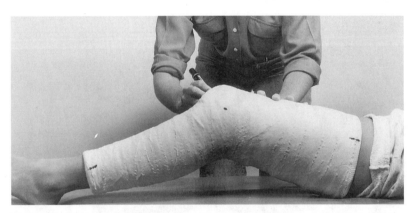

Figure 8–63 Mark medial and lateral condyle with dot.

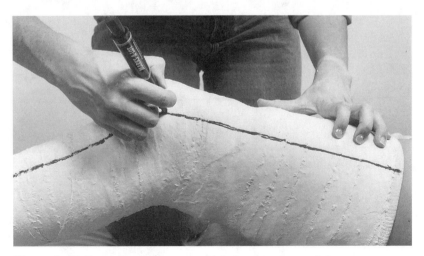

Figure 8–64 Draw line from upper thigh down to the condyle.

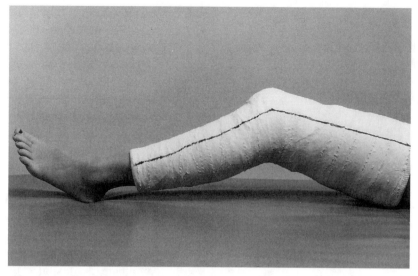

Figure 8–65 Continue drawing line down to ankle. Posterior shell should be two thirds in depth.

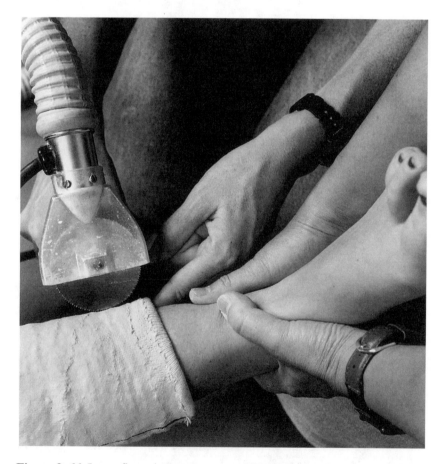

Figure 8–66 Insert finger in cast to move aside soft tissue. Cut along trimline.

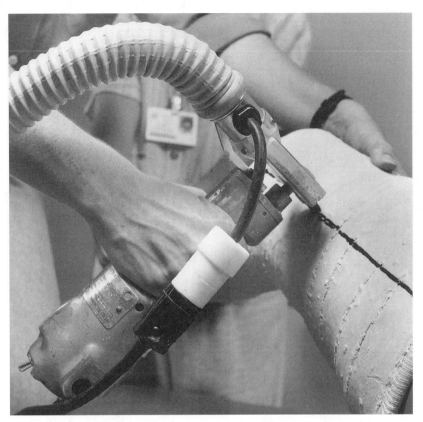

Figure 8–67 Cut along trimline.

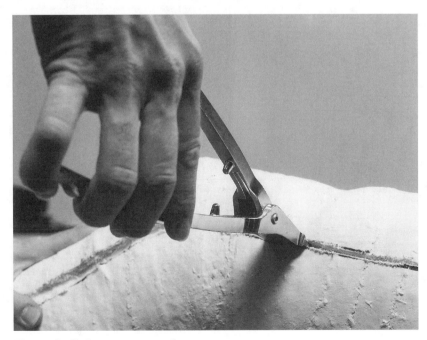

Figure 8–68 Insert cast spreader.

Figure 8–69 Separate the edges.

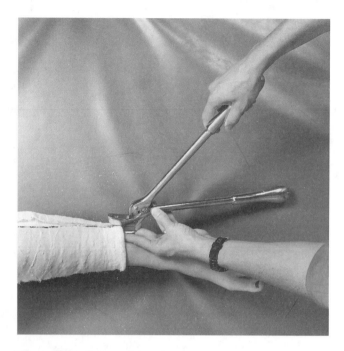

Figure 8–70 Insert crimper.

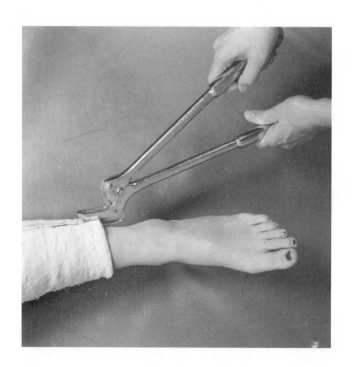

Figure 8–71 Cut stockinette with crimper.

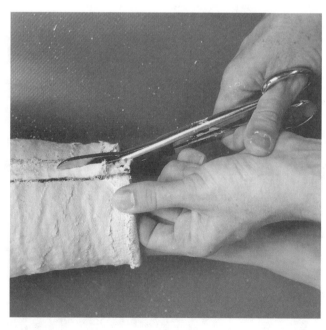

Figure 8–72 Cut stockinette and padding using bandage scissors.

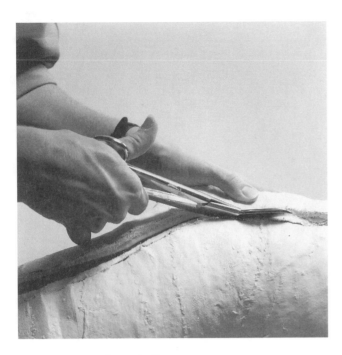

Figure 8–73 Follow trimline.

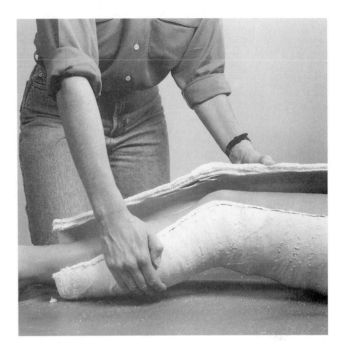

Figure 8–74 Remove anterior and posterior shells.

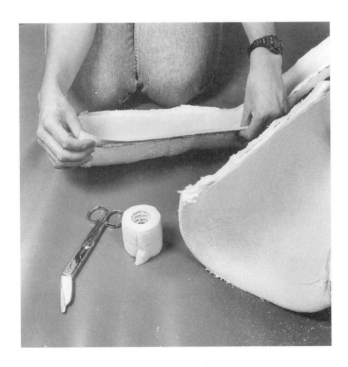

Figure 8–75 Apply tape to stockinette.

Figure 8–76 Secure stockinette to cast with tape.

CHAPTER 9

Considerations with Specific Diagnoses

Paula Goga-Eppenstein, Judy P. Hill, Terry Murphy Seifert, and Audrey M. Yasukawa

CEREBRAL PALSY

Early intervention is critical in optimizing overall function for the child with cerebral palsy (CP). The effects of abnormal tone and associated reactions that compromise the motor control needed for function and midline orientation present significant challenges to effective intervention.

Deformities caused by persistent asymmetry, postural malalignment, and compensatory movement may be prevented by very early treatment.[1,2] The occupational therapists involved in the treatment of this population employ various techniques to attain prevention of contractures and deformities, rebalancing of motor control, improvement of voluntary function, and maximal utilization of residual function. Clinical experience has shown that through careful assessment of the child with CP, upper extremity casting can be used successfully as an adjunctive therapeutic intervention focused on functional reeducation.

Casting an upper extremity is becoming common practice for the occupational therapist for use in managing abnormal muscle imbalance. The cast gradually elongates the involved muscle and soft tissues, allowing greater muscle length and rebalancing the opposing muscle group. Yasukawa[3] described the use of a long arm cast for a 15-month-old child with right hemiplegia. A muscle imbalance of the right arm limited active control of humeral external rotation, elbow extension, and forearm supination. A casting program was pro-

vided in addition to the weekly occupational therapy (OT) session. The focus of treatment was on strengthening the weak agonist muscles, which resulted in a better muscle balance around the joint, thus reducing the potential for myostatic contractures. Law et al.[4] described the effect of wrist casting and neurodevelopmental treatment (NDT) used either singularly or in combination on hand function in 73 children with spastic CP. The results suggested that casting combined with an NDT program improved the quality and range of upper extremity movement. The biomechanical alignment achieved through the use of casting provided joint stability with subsequent improvement of voluntary control and coordination. The researchers concluded that casting in combination with therapy appeared to facilitate a better response from a weak muscle group than therapy alone.

In 1993, Tona and Schneck[5] examined the effects of a short-term (48-hour) upper extremity cast with encased thermoplastic splint. The patient was an 8½-year-old girl with CP who had involvement in the left upper extremity and both lower extremities. The results suggested that the cast was effective in reducing spasticity and improving quality of movement. This case study suggested that temporary improvement in movement can be seen immediately after casting. However, to maintain lasting results the child must incorporate the movement into a functional skill. He or she must be intrinsically motivated to activate the weakened muscle. In addition, when the child is experiencing a growth spurt, the child must understand that the casting program may need to be reinstated to maintain the muscle length with the bone growth.

For children with myostatic contractures and poor control of the antagonist musculature, casting may lead to an increase in tendon length and therefore only have a transient effect on the contracture.[6,7] Cruickshank and O'Neill[8] described the effect of casting the upper extremities of an older child with spastic quadriplegia. Bivalve elbow casts were fabricated to improve range of motion into elbow extension. The results suggested that the upper extremity casts were effective in maintaining but not in improving passive range into elbow extension. Smith and Harris[9] also examined the effects of upper extremity casting in a child with spastic quadriplegia. They concluded that the purpose of casting was primarily to prevent further contractures and maintain range of motion in the elbows. Their results suggested that a positive effect of the casting program was ease of handling for the caregiver for positioning and hygiene care. Although casting may not lead to improvement in range or function in some cases, it may assist the physician and occupational therapist in the decision-making process of selecting a particular surgical procedure.

Assessment

When evaluating the upper extremity, one must look at the effect the trunk, pelvis, and lower extremities have on active reach. Ongoing assessment of the child's movement patterns during reach and fine motor activities is essential. The child with tonal problems generally develops compensatory movement patterns that eventually lead to the stronger muscle group's dominating the overstretched, weaker muscle group. A stable base of support and dynamic stability of the lower extremity and trunk are necessary if the upper extremity is to be free to function. Gentile[10] described the upper limbs and hands as having two modes of operation. In the first mode, the upper extremity is part of the overall postural system that is used in walking and running. The second mode incorporates the upper limbs with a stable base of support, such that the hands can engage in active manipulation for interacting with objects.

For children with CP it is extremely difficult to achieve this second mode since they cannot disengage

their upper extremities for functioning. They typically fix or stabilize with the upper extremity to assist with postural control in an upright position, lacking a stable base to work from otherwise.

A common problem seen with the upper extremity is the fixing pattern of scapular elevation, humeral extension, internal rotation, elbow flexion, forearm pronation, and wrist flexion, with thumb adducted or in the palm. This pattern is generally seen when there is an unstable base of support through the trunk or feet or a muscle imbalance in the affected arm. For example, humeral extension is often used to compensate for lack of thoracic extension. The hand and upper extremity may perform the weight shift that is normally accomplished by the lower trunk and pelvis.

When a therapist considers treatment options, selecting the type of cast requires a careful study of the involved arm or hand, in addition to an analysis of the compensatory patterns used in the entire body. Each child must be evaluated carefully to determine whether the goals of casting are (1) to facilitate or improve voluntary control and coordination, (2) to evaluate the potential muscle control to assist with the decision making for the therapy program or potential surgery, or (3) to prevent the development of further contractures or deformities.

Conclusion

The use of an upper extremity cast as an adjunct to a regular treatment regimen can be helpful when treating children with CP. It is not a treatment in itself, but should be used with more movement activity on the child's part. Casting the child at an earlier age in most cases will yield better results. This is because the child is probably not very tight or stiff and the deformities have not become established, so a greater potential exists to achieve muscle balance. The decisions about when to cast and for how long will depend on clinical judgment, observation, and careful assessment. Ongoing analysis is essential to ensure that the desired movements and quality of movements are achieved.

Case Study: C.M.

Initial Presentation

C.M. was 13 months old when she was initially evaluated by an occupational therapist, a physical therapist, and an infant development specialist. At 10 months, her mother had noticed that she was not using her left hand. A physician confirmed a diagnosis of CP with left hemiparesis for C.M.

In the initial assessment, an infant development specialist found C.M. to be functioning cognitively at her age level, although her vocabulary was somewhat limited. Her gross and fine motor skills were delayed, and she was at risk for developing orthopaedic problems. Her motor actions were clearly asymmetrical. She avoided weight shifting to the left and basically moved asymmetrically when prone and supine. When asked to reach for a toy in prone position, she reached with her uninvolved right upper extremity and retracted the left side of her body in an attempt to control the movement. The left lower extremity was generally postured in extension with the left upper extremity in a fixed or static position of humeral extension, internal rotation, elbow flexion, and forearm pronation with the hand fisted. C.M. was able to pivot in prone position but only to the right. She was able to maintain ring-sit briefly with a posterior pelvic tilt and trunk shortened on the involved side, such that her weight was shifted to the less involved side and the left arm held in a stiff posture to stay upright. In supported sitting with the trunk stabilized more symmetrically, she was able to swipe actively with the left arm and to use a gross grasp to reach for an object.

C.M. was seen weekly for a 1-hour OT session. The treatment focused on these four objectives: (1) to improve trunk and pelvic control on the involved side, (2) to improve lateral weight shifting in all positions, (3) to develop a stable base of support for facilitating greater left upper extremity movement, and (4) to decrease the compensatory fixing pattern of the left arm.

After 2 months of weekly therapy, C.M. was able to assume and maintain ring-sit independently from the right side only. The left trunk continued to be shortened with the left shoulder protracted, humerus extended, and distal arm held in a flexed position (Figure 9–1). A protective response of the left arm could be elicited, although delayed, with the hand fisted. She was able to assume and maintain sidelying independently onto the left side only (Figure 9–2). C.M.'s primary means of mobility developed as a modified combat crawl. During this movement her left arm was held fixed in shoulder elevation, humeral extension, and elbow flexion. She could reach forward with her uninvolved right arm and pull her entire body over the left shoulder. The left humerus was maintained in slight extension and internal rotation (Figure 9–3). In attempted forward reach in prone, the left arm did not move beyond a 90° range.

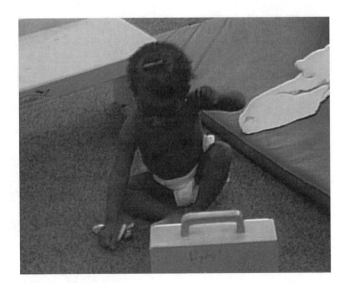

Figure 9–1 Sitting with left side of trunk shortened.

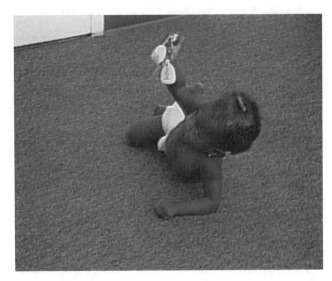

Figure 9–2 Sidelying on left side only.

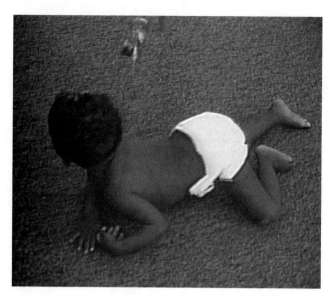

Figure 9–3 Crawling with left humerus maintained in slight extension.

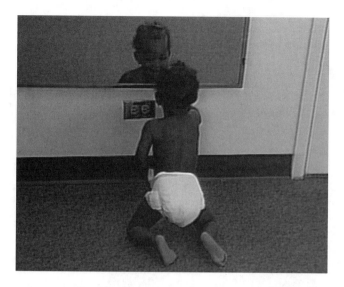

Figure 9–4 Semikneel with left arm held in a stiff position.

The latissimus dorsi, teres major, and pectoralis major were always held in a shortened position. There was also shortening of the anterior aspect of the glenohumeral joint capsule, thus limiting external rotation of the humerus. C.M. was able to assume and maintain a semikneeling position with both hips in flexion, the right upper extremity pulling the body erect, and the left arm held in a stiff position (Figure 9–4). Controlled movement of the left arm was limited in active range of external rotation and elbow extension secondary to the developing compensatory patterns. The stiffness in the left upper extremity inhibited her from developing the skills needed to assume and maintain quadruped for creeping.

Although the therapy focus was appropriate for decreasing the compensatory patterns and stiffness of the left arm, carryover after the session was minimal despite the family's efforts at home. Because of this, C.M. appeared to be a good candidate for casting. It was hoped that the use of a cast would help to facilitate greater left upper extremity arm movement and improve dissociation of the left arm from her trunk.

The Casting Program

A physician's order was obtained and a long arm cast was applied to C.M.'s left upper extremity, incorporating the elbow, wrist, and hand. This cast was positioned with the elbow extended and the forearm and wrist in neutral. The cast was kept on for a total of 10 days. While wearing the cast, C.M. was seen once for OT. The mother was instructed about activities for home to facilitate weight bearing on hands and knees,

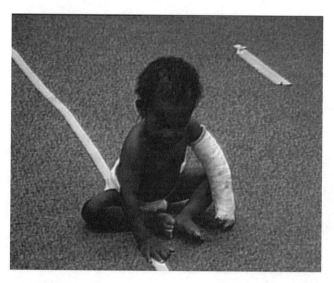

Figure 9–5 Sitting with cast.

Figure 9–6 Crawling with cast.

Figure 9–7 Sitting after casting.

creeping, and transitioning from sidelying or sitting to quadruped (Figures 9–5 and 9–6).

After the cast was removed, a bivalved long arm cast was fabricated to maintain the newly acquired length of elbow extension with the forearm slightly supinated. C.M. wore this bivalve cast every night and continued with the once-weekly outpatient OT sessions.

The immediate results of the 10 days of casting were increased proximal stability of the left shoulder and active triceps for weight bearing in quadruped, increased humeral flexion with adduction, improved scapular mobility, increased antigravity scapular control, dissociation between the scapula and humerus, and improved active reach overhead (Figures 9–7, 9–8, 9–9, and 9–10).

Figure 9–8 Transitioning into quadruped after casting

Figure 9–9 Crawling after casting.

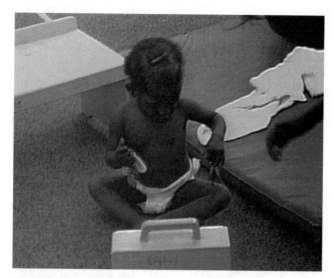

Figure 9–10 Left arm reaching after casting.

Henceforth the long arm cast was used during C.M.'s treatment sessions to facilitate the development of postural reactions and control, specifically in quadruped. The ability to weight bear on hands and knees, which is a critical stage for developing the mobility and dynamic stability of the hemiplegic arm, could be best achieved while wearing the cast. Furthermore, treatment in quadruped-facilitated palmar expansion and proprioceptive feedback were needed to enhance C.M.'s hand skills. Casting used as an adjunct to other treatment techniques provided the patient with the means to refine her stability in quadruped and other positions, thus improving her control to reach in wider ranges. The patient was aided in refining her quadruped position and improving active reach in wider ranges.

BRAIN INJURY

Individuals with brain injury are often candidates for casting due to neurologic damage which results in hypertonicity and loss of range of motion as well as prolonged periods of immobilization. This, in turn, leads to muscle tightness and loss of motion.

Studies have shown that a prolonged stretch is most effective in preventing and correcting contractures in connective tissue. Casting provides an effective and relatively inexpensive method to correct or prevent contractures that significantly limit an individual's functional level.

The clinician must perform a complete evaluation and consider what functional goals are to be achieved with casting. Along with this evaluation, the precautions and considerations previously discussed must be considered to determine if the individual will benefit from casting.

While casting, the clinician must be sure to continue to engage the individual in functional and therapeutic activities. That way, when an optimal range of motion is achieved, the individual will possess the motor skills, and have the ADL training, mobility and transfer training, to be able to use the increased motion function.

Once the casting process is complete, a plan must be put in place to assure that the individual will maintain range of motion. This must include patient, family, and caregiver education and may include bivalving, splinting, a strengthening program, a P/AROM program, and/or weight-bearing program. If the patient is permitted to go without follow-up, more than likely the increase in range of motion will be lost.

Upper Extremity Case Study

Initial Presentation

P.O. was a 23-year-old woman who was injured in a motor vehicle accident. She had a 3½-month acute hospitalization during which she was reportedly in a coma for 6 weeks. Lesions to the brain stem as well as diffuse cortical damage were identified on an electroencephalogram and a computed tomography scan. When admitted for rehabilitation, P.O. showed localized response to visual and auditory stimuli. She did not vocalize with the exception of loud moans occasionally, and this was combined with mass extensor response following imposed movement.

P.O. was markedly asymmetrical, with head rotated and laterally flexed to the left. The left upper extremity exhibited a strong flexor pattern and the right was in an extensor or decerebrate position. Some passive motion could be achieved with relaxation and positioning techniques, but severe deformity and hypertonicity were present bilaterally in the upper extremities. No spontaneous motion in the upper extremities was present with the exception of the mass extensor pattern noted previously, which involved the right upper extremity, neck, and trunk.

The lower extremities were generally less affected. Spontaneous, seemingly voluntary and controlled motion was present on the left. Minimal hip extensor and plantarflexion contractures were noted on the right.

The Casting Program

In initiating treatment, goals were to promote appropriate interaction with the environment by normalizing tone, correcting deformity, and mobilizing in the upper extremities using casting and neurophysiological treatment techniques. The program needed to be integrated to include general mobility and stimulation and make use of available lower extremity motion while managing the upper extremity abnormal tone and contractures.

Because of its apparent contribution to overall positional asymmetry, the right upper extremity was first casted into elbow flexion. This extremity was strongly patterned in the decerebrate position with

- scapular retraction
- shoulder extension
- internal rotation
- elbow extension
- forearm pronation
- wrist flexion
- ulnar deviation (Figure 9–11)

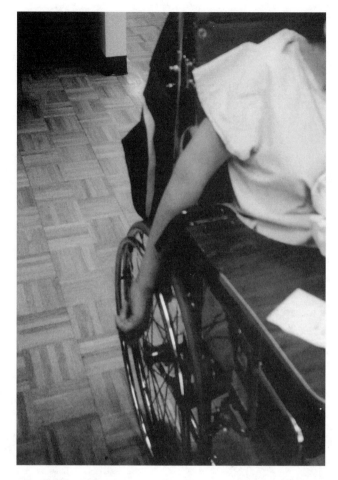

Figure 9–11 Right upper extremity/decerebrate position.

The fingers were contracted in flexion at the interphalangeal joints. While a rigid circular elbow cast could have been utilized to effect change at the elbow first, a long arm cast was selected because positioning the wrist in more extension and radial deviation and the forearm in supination helped in relaxing the tone throughout the extremity. By utilizing this method and incorporating shoulder flexion with the arm resting in front of the patient on a lap tray, the elbow could be flexed to 25°, and the forearm positioned in pronation rather than the usual hyperpronation. Following a series of two long arm casts, the elbow could be flexed to 100°, the wrist to 10° of flexion and neutral deviation (Figure 9–12). General posture, head positioning, and alertness were also improving. A bivalved cast was used at night for 8 days to help ensure that the gains were maintained. As overall tone was improving, a trial

Figure 9–12 Elbow flexion to 100°.

of having the patient use no night cast was introduced and range was maintained. At this point, for the right upper extremity, the focus was switched to the wrist and hand.

During the time period that the right bivalve was being used at night, intervention began to effect increases in elbow extension on the left side. This extremity exhibited moderated limitations throughout as follows:

- shoulder flexion to only 80°
- elbow extension to 80° of flexion (80° to 160°)
- wrist extension from 90° of flexion to 60° of flexion
- metacarpal phalangeal extension to a maximum of 20° to 30° of flexion
- proximal interphalangeal joint extension to a maximum of 50° of flexion

The usual resting position of the extremity was in approximately 100° of elbow flexion with wrist and fingers flexed (Figure 9–13). A series of two rigid circular elbow casts was applied and each was left in place for 5 days. Rigid circular casts were chosen over drop-out casts because of the extremity's fluctuating tone and positioning considerations. While the cast was in place, attempts to keep the hand from severe flexion included use of a cone and roll splint.

Following the use of the two casts, elbow extension had increased to lacking 30° from full extension, a gain of 50° (Figure 9–14). P.O.'s passive motion had increased, flexor tone had decreased, the resting position of the extremity was altered (Figure 9–15), and she was able to relax into extension on command. As on the right, it was decided to utilize a bivalve to see whether the elbow range could be maintained and then proceed with further casting for the wrist and hand.

To clarify how these casts were coordinated with each other, P.O. did not have two elbow casts on at the same time. The right long arm casts were used to increase flexion. While the night bivalve was being used to maintain improvements on the right, left elbow casting was initiated. While the second left rigid circular elbow cast was in place, wrist–hand casting began on the right to address the remaining flexion contractures (Figure 9–16).

Initially, the wrist was casted into extension with two casts. When the wrist could be extended to 30°,

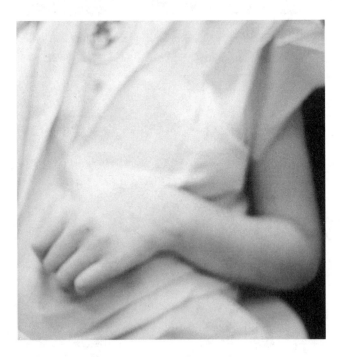

Figure 9–13 Resting position of extremity.

Figure 9–14 Increased elbow extension.

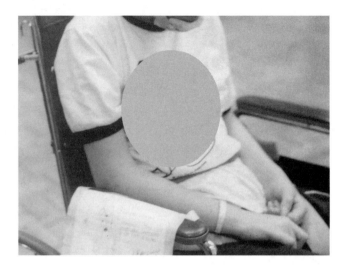

Figure 9–15 Altered position of extremities.

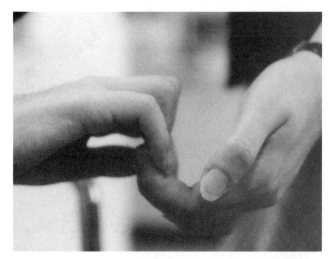

Figure 9–16 Remaining flexion contracture.

finger shell casts were added to increase finger extension of middle and ring fingers (Figure 9–17). After a total of four casts for the right wrist and hand, 45° of wrist extension could be achieved easily (Figure 9–18). With the fingers almost fully extended the wrist could easily be extended to 30° (Figure 9–19). Only minimal

volitional movement was noted in this extremity, however, even with the increases in range. A modified resting hand splint was made to maintain range.

The left wrist, which was more severely flexed, was casted during a period of a month with a series of five casts. This resulted in a gain of 65° of wrist extension,

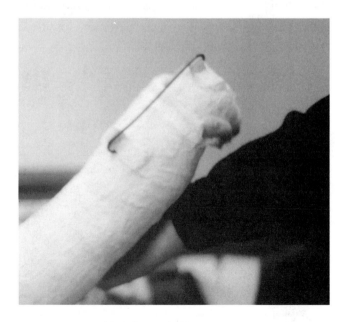

Figure 9–17 Added finger shell casts.

Figure 9–18 Wrist extension of 45°.

Figure 9–19 Wrist extension to 30°.

Figure 9–20 Gain of 65° of wrist extension.

from minus 50° to 15° of extension (Figure 9–20). The fingers could be partially extended both passively and to a lesser degree actively but were held in flexion most of the time. Outrigger extension assists were added to the next two casts to promote finger extension. Following these casts, beginning mass grasp and release were observed. The wrist could be extended to 60° and both metacarpophalangeals (MPs) and interphalangeals (IPs) could be extended to within 10° of normal range (Figures 9–21 and 9–22).

P.O.'s program focused on active management of contractures, while ensuring that one extremity remained uncasted to allow for spontaneous use as her arousal and interactions with the environment improved. While the casts were in place, P.O.'s extremities were actively mobilized and facilitated to engage in gross reaching and postural responses. The casting took place over 12 weeks and took parts of one or two sessions per week of therapy. Significant range of motion gains were achieved and some volitional motion

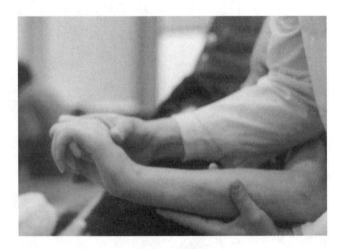

Figure 9–21 Wrist extension to 60°.

Figure 9–22 MPs and IPs extended to within 10° of normal range.

was revealed, more on the left side. In addition, P.O. began some functional arm placement to swipe at or reach for close objects and wipe her face.

Lower Extremity Case Study

Initial Presentation

L.M. was an 18-year-old woman who was involved in a motor vehicle accident with loss of consciousness. The initial CAT scan revealed a subarachnoid bleed. L.M. was measured as a six on the Glasgow Coma Scale given at the hospital. At discharge from the acute hospital, L.M. was following simple commands and purposely moving her left upper extremity. Prior to her accident, L.M. was a college student.

On the initial evaluation in rehabilitation, the patient was following three- to four-step commands, had decreased memory, and an attention span of approximately 5 minutes. The right upper extremity had minimum to moderate loss of motion; the left upper extremity was within functional limits; the right lower extremity was within functional limits with the exception of ankle dorsiflexion of negative 37°, eversion of 0°; the left lower extremity was within functional limits with the exception of dorsiflexion of negative 33°. Left upper and lower extremity strength was a good minus to good; right upper extremity mass flexion pattern; right lower extremity moved against gravity. L.M. required contact guard for bed mobility and moderate assistance for squat pivot transfers to all surfaces. L.M. was ataxic with all movements.

The Casting Program

The focus of L.M.'s treatment was mobility, transfers, W/C mobility, and balance training with serial casting to improve ankle/foot position for activities of daily living.

In 2 weeks, L.M. had received two sets of serial ankle casts with this resultant ankle range of motion: left ankle dorsiflexion — negative 13°; right ankle dorsiflexion — negative 8°. This was an increase of 29° on the right and 20° on the left. L.M. was now able to sit independently and accept challenges; she performed a squat pivot transfer with minimal assistance and performed sit to stand with minimum to moderate assistance.

In 4 weeks, L.M. was able to achieve neutral ankle dorsiflexion bilaterally in stance and was able to assume and maintain the position with supervision. L.M. was now able to ambulate 50 feet in the parallel bars with moderate assistance. She demonstrated decreased control with absent heel strike and initial contact on lateral borders of feet.

Casting continued and in 6 weeks L.M. presented with right ankle dorsiflexion of 0° to 2°; left ankle dorsiflexion of 0° to 8°. The patient was now performing more standing and ambulation, and further casting and splinting were not required. However, the patient continued to receive daily ankle stretching and mobilization.

In 8 weeks, the patient was able to maintain stance with supervision with upper extremity support. She was able to perform a squat pivot transfer independently and was able to ambulate 60 feet with a large base quad cane and minimal to moderate assistance.

At discharge, 11 weeks post-admission, L.M. demonstrated good safety awareness and was able to recall previous activities. Ankle range of motion was left ankle dorsiflexion 0° to 6° and right ankle dorsiflexion 0° to 5°. L.M. had active movement at all joints in the right lower extremity, but ataxia remained a problem. She was able to stand with upper extremity support and supervision. She was able to ambulate 150 feet with a walker and supervision and 80 to 100 feet with a large base quad cane and minimum assistance. She had increased dorsiflexion bilaterally at heel strike; however, ataxia limited her independence with ambulation. L.M. was able to climb stairs with a small base quad cane and contact guard to minimal assistance. L.M. was discharged home with continued therapy on an outpatient basis.

On recheck several years post-discharge, L.M. had graduated from college and was ambulating independently with a straight cane. L.M. continued to display minimum ataxia with open chain movement, but it did not interfere with her everyday functioning.

Through serial ankle casting, L.M. was able to achieve foot flat for more stable sitting, standing, and ambulation. Without that improvement of 40° plus of ankle motion, L.M. would not have progressed with her higher level mobility skills.

Lower Extremity Case Study—Knee

Initial Presentation

A.M. was a 52-year-old man who fell 50 feet and sustained a traumatic brain injury and multiple fractures including bilateral humeral head; right radius; C 7 and T 1; left superior ramus; and bilateral shoulder dislocation. CAT scan revealed a subarachnoid hemorrhage with bilateral frontal/temporal contusions. A.M. was admitted to rehabilitation approximately 4 months post-injury.

Upon admission A.M. was alert, oriented to self with poor attention, safety awareness, and judgment. He had significant range of motion limitations in bilateral upper extremities, his right lower extremity was grossly within functional limits, his left lower extremity was negative 25° hip extension, 0° internal rotation, and negative 40° knee extension. A.M. had no active movement in his left upper extremity, but had active elbow and wrist movement in his right upper extremity. He had active hip and knee flexion in bilateral lower extremities. A.M. presented with moderate increase in flexor tone in bilateral upper extremities and flexor withdrawal in bilateral lower extremities. A.M. was dependent with all mobility skills.

The Casting Program

It was determined that A.M. would benefit from serial casting to increase left knee extension to allow for improved weight bearing. A fiberglass cast was placed on the left lower extremity at negative 44° extension. Care was taken to pad the proximal and distal ends of the cast to avoid skin breakdown. The cast remained in place for 7 days and was removed with a gain in range of 10° of extension. A.M. was re-casted at negative 33° left knee extension.

Two weeks post-admission, A.M. had made minimal gains in mobility. He was able to perform bed mobility with moderate assistance, sit unsupported for short periods with supervision and cues, and perform transfers to various surfaces with moderate to maximal assistance.

Casting continued and 4 weeks post-admission the left knee extension range had improved from negative 33° to negative 20°. A.M. was now able to perform transfers to varied surfaces with moderate assistance

consistently. A.M. was able to stand and ambulate 5 feet with moderate assistance and maximal cues. Limited knee extension continued to limit A.M.'s ability to stand.

Six weeks post-admission with continued serial casting, left knee extension range had improved from negative 20° to negative 8°. A.M. was now able to propel his wheelchair with minimal assistance and maximal cues. A.M. remained able to ambulate 5 feet in the parallel bars with moderate assistance. Progress with ambulation was limited due to limited left knee extension and limited left ankle dorsiflexion.

Casting to left knee continued and 8 weeks post-admission, A.M. achieved 0° left knee extension. During A.M.'s last 2 weeks of inpatient rehabilitation, he was transferred to acute care, and a right shoulder hemiarthroplasty was performed. Due to this complication, casting was not initiated on the left ankle to improve dorsiflexion range. It was decided that a nerve block to the left gastroc-soleus complex followed by casting would be initiated on an outpatient basis.

At discharge, A.M. was cooperative in therapy and was able to express needs. His ability to attend to a task and follow commands remained a limiting factor. A.M. was able to perform bed mobility with minimal assistance and transfer with moderate assistance. He was able to maintain sitting with supervision and accept minimal challenges. Standing continued to require moderate assistance; however, he had improved left hip and knee alignment.

With serial casting, A.M. was able to achieve full left knee extension. This ability to extend his left knee allowed for more normal hip position in standing. A.M. was followed by a different facility than ours for outpatient services. With continued casting to improve left ankle dorsiflexion, A.M. has a greater chance for functional ambulation as this will allow for heel strike and foot flat stance phase. We hope that A.M. received continued casting and achieved his goal of ambulating.

Lower Extremity Case Study—Ankle

Initial Presentation

D.S. was a 34-year-old woman with a diagnosis of traumatic brain injury secondary to a motor vehicle accident. CAT scan revealed bilateral anterior parietal contusions, right focal cerebellar hemorrhage, and right parietal contusion. Upon admission, D.S. was alert and oriented times three. She was impulsive with decreased problem solving and carryover. D.S. presented with left ankle dorsiflexion of negative 25° and right ankle dorsiflexion of negative 35°. She was able to move her left lower extremity actively in all planes; however, movements were slow and uncoordinated. She was able to move her right lower extremity out of synergy in gross flexion or extension movements. D.S. was able to sit independently and stand with bilateral upper extremity support and minimal assistance. Positive support reflex was noted bilaterally, with weight bearing occurring through the fourth and fifth metatarsal heads. She required moderate assistance for transition movements and lateral transfers.

The Casting Program

It was determined that D.S. would benefit from bilateral ankle serial casting to improve ankle dorsiflexion for improved weight bearing in transfers and standing. Two weeks post-admission, one set of serial casts had been applied with an increase of 15° in right ankle dorsiflexion and 13° left ankle dorsiflexion post weight bearing.

Four weeks post-admission, ankle dorsiflexion remained right: negative 18° and left: negative 10°. Due to the need to work with D.S. in standing on balance and pre-gait activities, Aquaplast splints were fabricated with tone-reducing footplates to be worn at night and during the day when she was in her wheelchair. Bivalve ankle splints with posting at the heel were made so that D.S. would weight bear through her entire foot and minimize the positive support reflex. At this point, her mobility had improved with independent bed mobility and minimal assistance with lateral transfers. It was decided that D.S. would be more independent with transfers if she used a sliding board secondary to positive support reflex interfering with her ability to bring her weight forward over her feet.

During the next interim, the physician agreed to perform marcaine injections to bilateral Achilles tendons just prior to casting. Post-injection, D.S. was placed in prone position to decrease the influence of extensor tone and allow for increased force into dorsiflexion

while holding. The cast was applied and posted with dicem applied to the base of the cast to allow her to stand in casts with full contact across her feet.

After removal of the casts, right ankle dorsiflexion increased from negative 12° to positive 1° and left ankle dorsiflexion increased from negative 15° to negative 2°. Aquaplast splints were then fabricated for night wear and she tolerated them. Functionally, she was able to perform a sliding board transfer with contact guard to supervision. In standing with an orthotic; she required moderate assistance to maintain, and initially ankle dorsiflexion was negative 4° bilaterally. With prolonged weight bearing, she was able to achieve foot flat; however, as D.S. would unweight to step, her positive support reflex would fire. To decrease the influence of this reflex and allow for gait training, supramalleolar orthoses (SMOs) with tone-reducing footplates were fabricated with fiberglass. The cast was bivalved, a toe break was incorporated to allow for toe off in final stance phase, and dicem was applied to the bottom of the casts to avoid slipping.

D.S.'s ankle range continued to improve with increased time in weight bearing. Right ankle dorsiflexion increased from positive 1° to positive 5° and left ankle dorsiflexion increased from negative 2° to positive 2°. She began work in gait training and was taking 5 to 10 steps forward and back with tactile cues for anterior weight shift and upright posture. Her balance was improving with the ability to accept minimal challenges.

Focus of treatment continued to be increasing independence with all mobility skills. As discharge approached, the orthotics department was contacted and bilateral SMOs with inhibitory footplates were fabricated from thermal plastic to wear inside shoes. At discharge, D.S. had maintained ankle dorsiflexion range. She was independent with all bed mobility; she was able to perform sit to stand with contact guard and maintain standing with SMOs and contact guard; she was now independent with all sliding board transfers. D.S. was able to ambulate with bilateral SMOs and rolling walker 100 feet with contact guard. Without SMOs, she required maximal assistance to ambulate with rolling walker. D.S. was able to climb stairs with SMOs, small base quad cane, and moderate assistance. She was discharged home with continued therapy scheduled on an in-home basis secondary to difficulty with transportation.

Three years following discharge from acute rehabilitation, D.S. returned for a recheck visit. She had maintained bilateral ankle range of motion and was now ambulating independently with a straight cane. D.S. continued with some gait deviations and a home program was issued in attempts to improve this, but her ambulation was functional and allowed her independence in the home and short distances in the community.

Through multiple casting and splinting interventions, D.S. was able to gain the mobility necessary to achieve foot flat, which significantly impacted her ability to transfer and ambulate.

SPINAL CORD INJURY

Upper Extremity Case Study

Initial Presentation

B.C. was a 24-year-old man who had sustained a cervical spinal cord injury 3 years previously that resulted in C5 quadriplegia. At admission, he presented with elbow flexion and supination and wrist extension contractures bilaterally (Figure 9–23). On quick stretch to the biceps, there was a mildly exaggerated response near the end of the available range. Utilizing his available 50° arc of elbow flexion, B.C. could bring his supinated palm to his mouth and then relax his biceps back to his usual resting position at 95° of elbow flexion.

Neither upper extremity was functional for any basic activities of daily living. Other than the partial hand-to-mouth pattern, no functional arm placement patterns were available. Passively, B.C.'s elbows could be stretched to 50° of flexion and 45° of pronation. Maximum pronation where the elbow was extended to 50° was 20°. Maximum wrist flexion was 15° of extension. Serial casting was initiated, first on the dominant right side with the goals of achieving independence with equipment and setup in eating and keyboard use.

The Casting Program

A long arm cast was applied with the elbow at 60° of flexion and the forearm at approximately 20° of pronation (Figure 9–24). When the first cast was removed

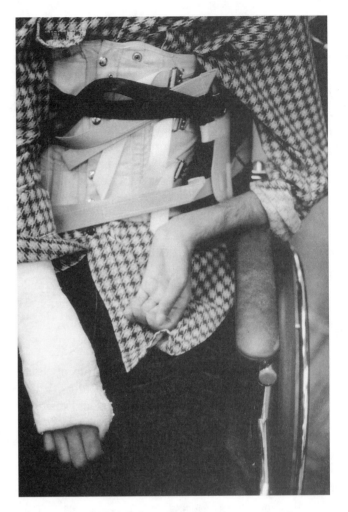

Figure 9–23 Elbow flexion, supination, and wrist contractures.

after 5 days, B.C.'s elbow could be stretched to 30° of flexion and forearm pronation remained 45° (Figure 9–25). B.C. could bring his hand to his mouth with his forearm slightly pronated so that his thumb would touch his mouth. He had difficulty maintaining this position, however, and his forearm tended to turn back into supination. A second cast was applied with the elbow extended to 35° of flexion and forearm in 40° pronation (Figure 9–26). This cast was also left in place for 5 days. When it was removed, passive elbow extension was 20° flexion and pronation was 65°. Wrist flexion to neutral was possible to this point. B.C. was able to bring his hand to his mouth while maintaining pronation and rest his arm on a table or lap tray in pronation

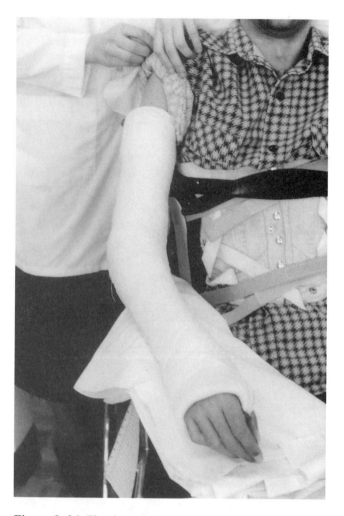

Figure 9–24 First long arm cast.

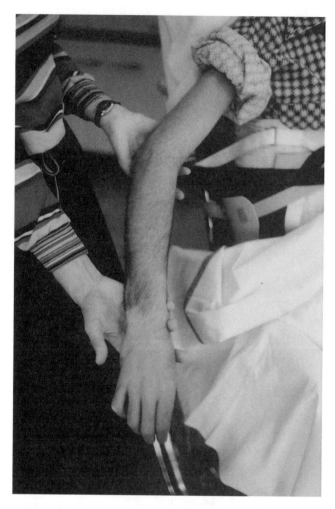

Figure 9–25 Results after first cast removed.

(Figures 9–27 and 9–28). He was also able to begin feeding himself and reaching a keyboard for computer access with a utensil holder in a long opponens wrist–hand orthosis. B.C. required a bivalved cast at night to guard against recurrence of the contractures. He also would have been a good candidate for tendon transfer surgery to provide active elbow extension and pronation.

As with many individuals with spinal cord injuries at the C5 level, range of motion can frequently be improved with casting[11] enough to result in functional improvements, but with counterbalancing musculature not innervated, follow-up positioning is almost always necessary. While a night use bivalved cast can maintain the range, it is heavy and often interferes with comfortable positioning. Surgical intervention[12,13] to provide a muscular counterbalance to the biceps should be considered for long-term maintenance and management.

BURNS

The care of acute burn injuries may consist of surgery followed by an intensive regimen of positioning, splinting, hydrotherapy, exercise, and pressure garments. The nature of the burn, the location and percentage of the body surface that has been burned, and the depth and degree of injury sustained require special consideration. When developing a treatment program,

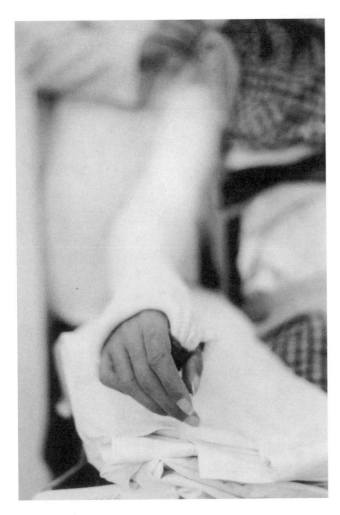

Figure 9–26 Second long arm cast.

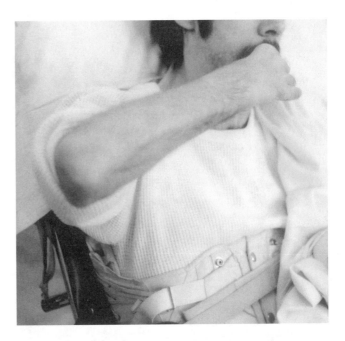

Figure 9–27 Hand to mouth while maintaining pronation.

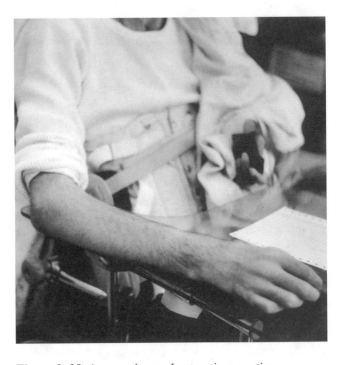

Figure 9–28 Arm resting on lap tray in pronation.

cooperation with the patient, family, and burn team is critical for promoting healing.

Burn scar tissues have a tendency to lose elasticity and become hypertrophic. The use of pressure garments has been known to influence and prevent hypertrophic scarring.[14] When a hypertrophic scar has been left unresolved and occurs across a joint, a contracture may result that eventually will limit active and passive range of motion. More important, serious contractures may limit the patient's independence in functional ability.

Contractures can be successfully prevented by using a combination of positioning, splints, and exercise. In the pediatric and adolescent patient populations contractures are especially problematic, not because of the severity of the burns but frequently because of the poor compliance of these patients in carrying out the rehabilitation regimen.[15] Serial plaster casting can be an effective treatment method for reducing contractures in the patient who does not comply with splinting, exercise, and positioning.

Plaster casts provide gentle, gradual tissue lengthening and appear to soften the hypertrophic scar (Figures 9–29, 9–30, 9–31, and 9–32). Bell-Krotoski and Figarola[16] state that by using the method of plaster casting, time appears to influence the growth of tissue. The constantly applied tension for an extended time interval

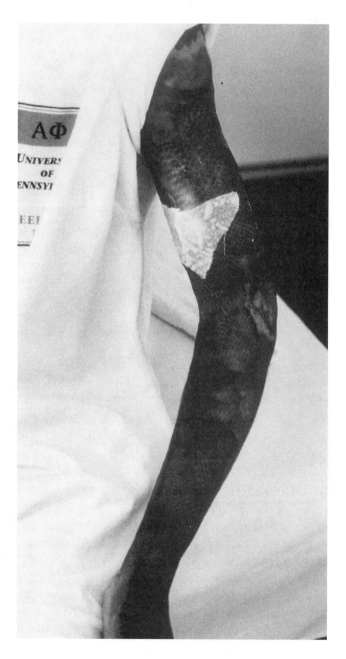

Figure 9–29 Contracted elbow.

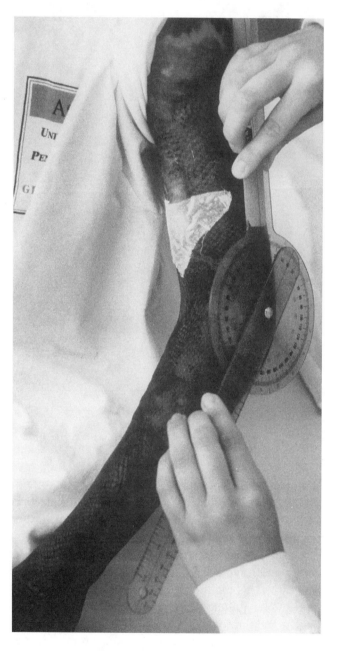

Figure 9–30 Limitation into elbow extension.

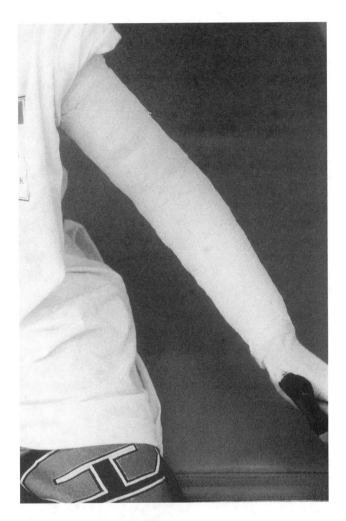

Figure 9–31 Rigid circular elbow cast.

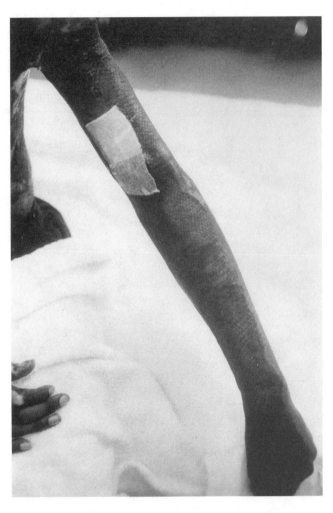

Figure 9–32 Full range of motion into elbow extension.

promotes remolding and growth of the skin and soft tissue. To maximize results, a plaster cast is left on for 3 to 7 days until either full range returns or no further gains are achieved.

The use of serial casting with burns can be used to correct a contracture with active exercise and daily routine care. This requires ongoing evaluation during the patient's rehabilitation and follow-up. Education of the patient and family is one of the most important factors required to prevent contractures and promote functional mobility and strength.

JUVENILE RHEUMATOID ARTHRITIS

Juvenile rheumatoid arthritis (JRA) is a chronic disease, which upon onset may be characterized by inflammation, pain, and limitation of motion in one or more joints for a period of 6 weeks or more. The history of onset cannot be attributed to any known cause such as trauma, rheumatic fever, or infection.[17,18] To be diagnosed with JRA the child must be under age 16. In addition, a segment of the JRA population will test with a rheumatic factor (RF) in their blood. Children with a seronegative RF tend to have milder arthritis with a greater remission rate than those with a positive RF.

During the first 6 months after onset, certain unique patterns of joint involvement, type of onset, and symptomatology exist that further subdivide the group.[17,18] The three major subtypes include the following characteristics:

- Systemic onset—high spiking fever, rheumatoid rash, variable joint involvement, negative RF, and possible involvement of other organ systems
- Pauciarticular onset—involvement of four or fewer joints, negative RF
- Polyarticular onset—involvement of five or more joints, negative and/or positive RF

Only the polyarticular onset with positive RF resembles adult RA and may result in the greatest joint destruction.

Common to all subgroups is joint inflammation, chronic synovitis, general fatigue, and muscle weakness. Inflammation begins in the synovial membrane that lines the joint capsule. It contributes to an altered level of joint fluid, which in turn results in the destruction of the joint capsule, supporting ligaments, cartilage, and finally bone. As a consequence, this progressive, chronic disease affects the upper extremities by deforming joints and damaging soft tissues.

Chaplin et al.[19] described 414 patients with JRA, 66% of whom showed involvement of the joints of the wrist and fingers. The common abnormalities of the hand included the lack of wrist extension, wrist ulnar deviation, and finger abnormalities, specifically swanneck or boutonniere deformity. Findley et al.[20] stated that wrist involvement is seen in the majority of JRA patients' disease process. In describing the stability of the carpal bones, Findley et al. stated that the dorsal and ulnar collateral ligaments are much thinner and weaker than the volar and radial collateral ligaments, contributing to the commonly seen palmar subluxation of the wrist. In the patient with more severe JRA, ankylosis of the wrist was seen. Athreya[21] discussed problems with laxity of ligaments around the metacarpophalangeal (MCP) joint of the thumb. He discussed three stages of thumb involvement: (1) laxity of ligaments and volar subluxation, (2) rotation of the thumb so that it takes the same axis as the fingers, and (3) gradual loss of the

web space. Nalebuff[22] developed a classification system for the more common thumb deformities.

Clinically, children with JRA display decreased muscle strength, pain, and limited range and function. The most critical factors in long-term care are good control of the inflammation and preservation of joint function. The rheumatologist's and the primary physician's roles are to consider the course of drug therapy and monitor the status of the child and the response to the selected drug. The occupational therapist works with the physician to prevent long-term disability and loss of function.

Casting

Inflammation and synovitis of the joints or tendons in the hand are common manifestations of the disease and are the primary mechanism of decreased function or destruction of the joint.

Casting methods can be used successfully as an adjunct to the therapy program in treating upper extremity contractures other than those due to a bony restriction. A radiograph is necessary to rule out any contraindication. In children, however, cartilage is not visible on the x-ray film, creating an impression of joint space. Melvin states, "the term joint space means cartilage in reference to a radiograph."[17(p146)] In very young children, cartilage damage will not show until they reach skeletal maturity. However, in older children damage or destruction of the joint is clearly evident.

The rationale for casting includes (1) prevention and correction of the deformity, (2) increase in function, (3) control of inflammation and preservation of the joint, (4) relief of pain from muscle spasm, and (5) restoration of strength. Casting can assist with remolding and contouring the joint and muscle to increase range of motion by providing a gentle gradual stretch and positioning the tissues near the end range of their elastic limits.

Contraindications to casting may include an unstable, painful joint in which increasing the range may also increase the pain. Casting may not improve and/or may cause further irritation and pain for joint surfaces made irregular from erosion and swelling.

Sometimes, local injection of the wrist with corticosteroids is performed in some patients to control the inflammation and relieve the muscular spasm affecting the joint. The use of a wrist cast in conjunction with corticosteroid treatment may help keep the joint quiet, provide joint support, decrease the pain and spasms, and improve range.

Cast intervention with JRA patients must be carefully timed and based on the inflammatory activity phase. Generally, the goal during the acute phase is to rest the joint, position the hand in a functional position, and decrease the inflammation. If a cast is to be used, it should be on for approximately 24 hours with a follow-up hand splint.

When the pain and inflammation have significantly subsided, a serial casting program can be implemented. The cast must be changed every 48 to 72 hours. In the subacute stage there is less pain, and casting can be used. Painful muscle action and spasms may be decreased by aligning the joint in a comfortable yet improved biomechanical alignment. In the chronic phase, more aggressive corrective measures may be needed to prevent or correct the contractures.

In JRA active and passive range of motion of the involved joints must be performed frequently to prevent loss of functional range. For this reason, casts are changed more frequently than with other diagnoses. In all cases aggressive treatment is essential to prevent contractures and minimize the muscle imbalance and biomechanical malalignment.

Conclusion

Casting the JRA patient requires careful observation and appropriate intervention throughout the disease process. The goals for casting are related to the phase of the inflammatory activity. The long-term goal is to maintain or improve range and functional strength and slow the deterioration that typically occurs with time.

Case Study: P.K.

Initial Presentation

P.K. was a 14-year-old girl with progressive polyarticular JRA who was RF positive. The onset of the disease occurred when she was age 7. P.K. complained of pain in her right hand that was greater than in her left hand. She exhibited slight synovitis of the right wrist and tenderness in the MCP joint of the third digit. In addition, she reported right shoulder pain with forced abduction and external rotation.

P.K. stated that she was losing range and strength in her right wrist. She was having difficulty keeping up with her classmates in her typing class. Passive range measurements revealed the following:

Wrist	Left	Right
Extension	0°–60°	0°
Flexion	0°–70°	0°–60°
Radial deviation	0°–25°	0°
Ulnar deviation	0°–30°	0°–25°

X-rays of the right hand revealed fusion of the hamate and capitate. Soft tissue swelling was overlying all of the proximal interphalangeals (PIPs) and a small erosion existed at the base of the third PIP. There were erosions of the distal radius and ulna, as well as the carpal bones, and narrowing of the radiocarpal and carpometacarpal joints. These abnormalities had increased since the previous exam.

The chronic synovitis of the right wrist joint weakened the supporting ligaments about the wrist. As a result, there was volar slippage of the carpal bone on the radius. This subluxation further resulted in extensor carpi ulnaris (ECU) tendon slippage, which created an additional flexor force. Pain and spasm of the flexor carpi ulnaris (FCU) occurred with the wrist posture in slight flexion and ulnar deviation. The ECU acted as an ulnar deviator (Figures 9–33 and 9–34).

The Casting Program

A wrist cast was applied with emphasis on supporting the proximal carpal bone to reduce the deforming tendency at the wrist joint (Figure 9–35). Accuracy was important in the pressure, counterpressure, and traction applied so that the proper torque was created to position and align the wrist joint. The cast was extended to the midpalmar crease to provide free finger and thumb movement. With the cast in place, P.K. was encouraged to perform isometric contraction of wrist extension.

The wrist cast was changed each day. After the first cast was removed, P.K. stated that her wrist looked and felt different. Fluidotherapy and passive and active range of motion exercises were initiated on her right wrist following each cast change. The initial gain of

Figure 9–33 Right wrist subluxation.

Figure 9–34 ECU acted as an ulnar deviator.

range was from 0° to 15° extension (Figures 9–36 and 9–37). She had a series of four casts applied with an increase in range of motion from each cast change (Figures 9–38 and 9–39). When the final cast was removed, she had a gain in range of motion from 0° to 50°. A follow-up orthosis was fabricated to maintain the range of motion gain from the casting program, and an aggressive exercise program was implemented to strengthen and restore muscle balance. P.K.'s exercise program goals consisted of improving active wrist extension and radial deviation as well as strengthening the ECU (Figure 9–40).

P.K. has been coping quite well with her arthritis, although she still has approximately 30 minutes of morning stiffness. With weather changes or during an increase in her activity level, she reports joint pain. She continues to take her daily medications to control inflammation and monitors how much to push her body.

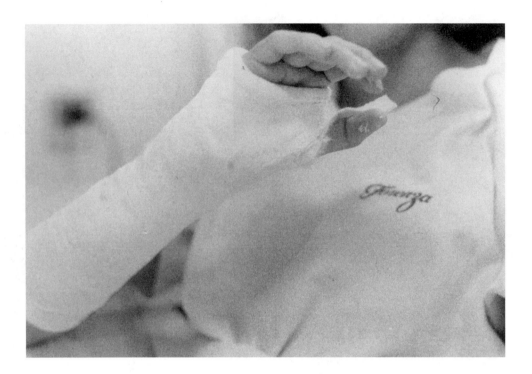

Figure 9–35 Right wrist cast.

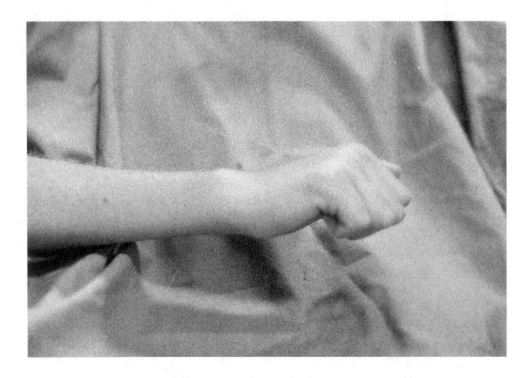

Figure 9–36 Wrist extensor 0° to 15° with fingers flexed.

Figure 9–37 Active extension of wrist and fingers at 0° to 15°.

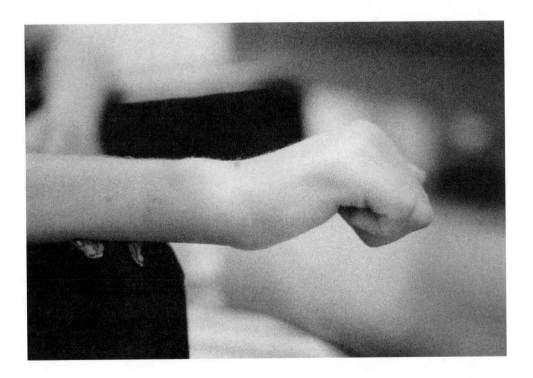

Figure 9–38 Continued improvement into wrist extension.

Figure 9–39 Continued improvement of wrist and finger extension.

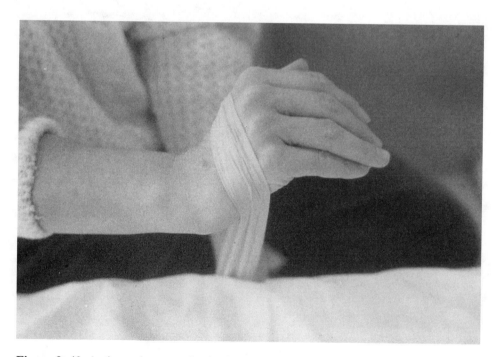

Figure 9–40 Active wrist-strengthening program.

P.K. has maintained the range in her right wrist for the past year. She wears a right wrist orthosis at night. She types her homework on a computer or electric typewriter, and her only limitation in typing class is decreased speed. She continues to use joint protection techniques and maintain correct postural alignment while performing tasks in school.

Case Study: H.N.

Initial Presentation

H.N. was a 7-year-old boy who was diagnosed with polyarticular RF-negative JRA at the age of 4½ years. Since his diagnosis, he has been treated with a variety of medications with only a fair response. Despite frequent occupational therapy sessions and excellent compliance with a home program, H.N. continued to lose range of motion in all his joints. By radiograph it was noted that he had significant erosive changes that were in his hands, wrists, hips, and knees.

H.N. demonstrated difficulty performing his daily functional care. He had weak hands bilaterally and was unable to unscrew the toothpaste cap, snap and unsnap, and button and unbutton his pants or shirt. His poor grasp and pinch limited his ability to perform fine motor skill activities, as well as his activities of daily living.

Clinically, H.N. exhibited the Nalebuff's classification[22] type 1 pattern, the extrinsic minus thumb. His deformity consisted of swelling secondary to synovitis of the MCP joints bilaterally in the thumbs. The extensor pollicis brevis (EPB) became overstretched, as well as the joint capsule and collateral ligaments. The extensor pollicis longus (EPL) slipped volarly, causing the MCP joint to flex and the IP joint to hyperextend (Figure 9–41). In addition, there was spasm and tightness of the intrinsic thumb muscles causing an intrinsic plus position of the MCP in flexion and IP in hyperextension (Figures 9–42 and 9–43).

The range of motion measurements of his wrist and thumb revealed the following:

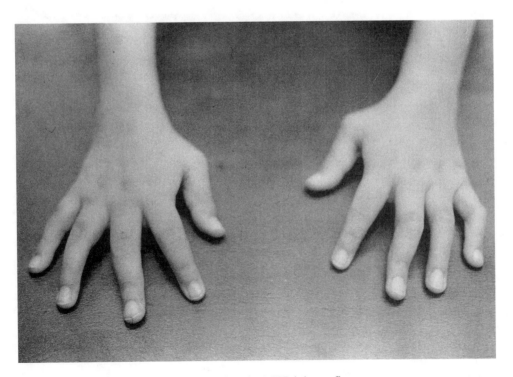

Figure 9–41 EPL slipped volarly, causing the MCP joint to flex.

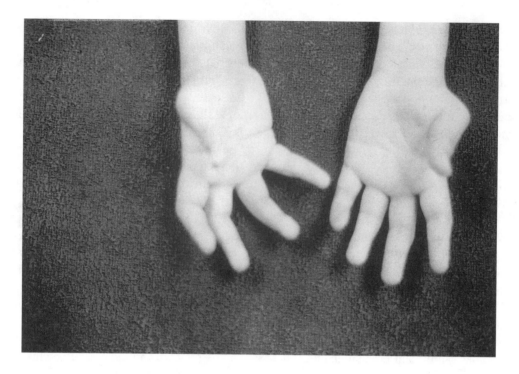

Figure 9–42 Spasm and tightness of the intrinsic thumb muscles, causing an intrinsic plus position.

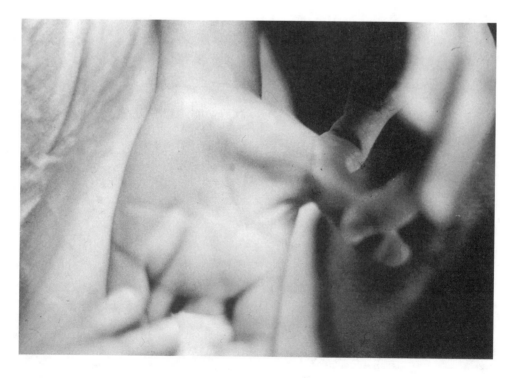

Figure 9–43 Spasm of the thumb flexor.

Wrist	Left	Right
Extension	0°–45°	0°–45°
Flexion	0°–80°	0°–80°
Ulnar deviation	0°–30°	0°–30°
Radial deviation	0°	0°

Thumb	Left	Right
Flexion MP	35°–50°	35°–50°
Flexion IP	–10°–70°	–10°–70°
Abduction	0°–60°	0°–60°

The Casting Program

A series of wrist casts with the thumb enclosed was applied to the left hand to realign the thumb MCP joint and gradually stretch the tight intrinsic thumb musculature (Figure 9–44). The first cast was removed within 24 hours. Initially, the left hand was casted and the right hand was used to compare as a baseline for change (Figure 9–45). Of concern was the potential of losing range in his wrist due to immobilization in the cast. Upon removal of the cast, paraffin was used to decrease the stiffness in his hand, as was passive and active range of motion. H.N. continued to maintain his initial range measurements at the wrist. Improvements were noted in the MCP alignment, as well as a decrease in tightness in the thenar eminence (Figure 9–46).

H.N. required a series of five wrist casts with the thumb enclosed; following removal of the first cast in 24 hours, each cast was on for a 48-hour period. In between each cast change, range measurements were taken and skin checked. In addition, an aggressive treatment program, which consisted of paraffin treatment and active/passive range of motion exercises, was given before reapplying the next cast. H.N. attained full passive range of motion in his left thumb and continued to maintain the range in his wrist (Figure 9–47).

A follow-up night resting splint was fabricated to maintain the alignment of the thumb MCP joint in extension and abduction. During the day a soft neoprene thumb glove was worn to support and stabilize the MCP joint. For the next 5 months a functional electrical stimulation (FES) unit was rented for his daily use at

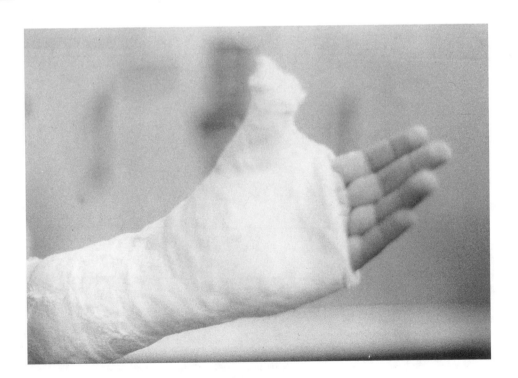

Figure 9–44 Left wrist cast with thumb enclosed.

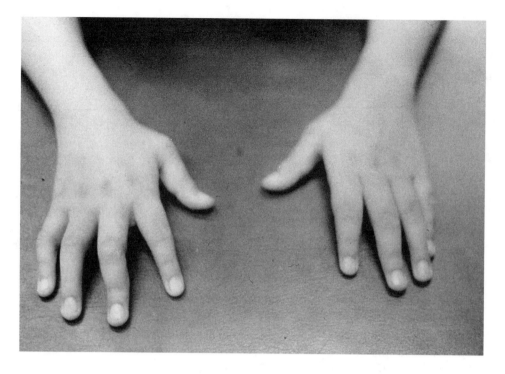

Figure 9–45 Comparing left casted wrist and thumb with right noncasted hand.

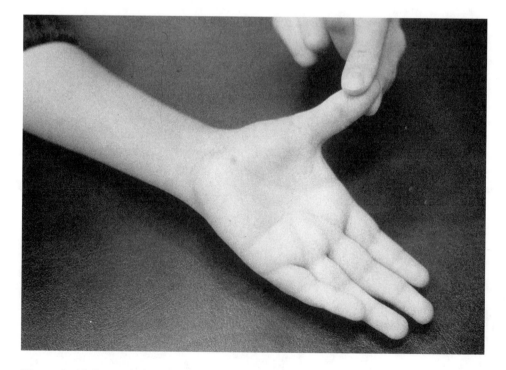

Figure 9–46 Decreased tightness of the thenar eminence.

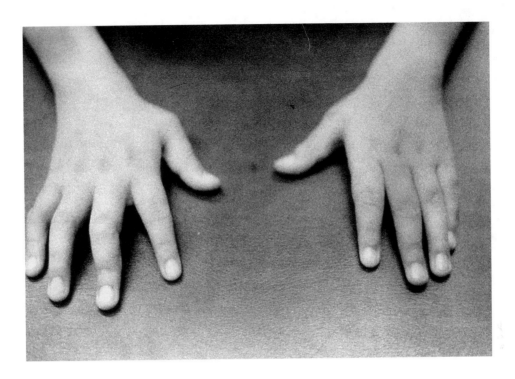

Figure 9–47 Completion of left wrist with thumb-enclosed casting series.

home. The FES unit was used three times per day to reeducate and strengthen the thumb extensors.

H.N. later developed a boutonniere deformity in the fingers of both hands. This deformity consisted of flexion at the PIP joints and hyperextension at the distal interphalangeal (DIP) joints. His mother was instructed about how to apply the finger cast using the two-stage technique.[23] The plaster bandage was cut into 1-inch strips and applied initially on the DIP joint with the PIP joint positioned in flexion to reduce the pull of the lateral bands. The DIP joint was positioned into flexion during cast application and held until the cast was hardened. Next the PIP joint was extended and the second cast was applied beginning at the MCP crease. His mother changed the finger casts every other day until full range of motion was attained. In addition, H.N. was being seen two times per week by his occupational therapist.

Functionally, as H.N.'s hands became stronger, he developed independence in his self-care tasks. He was able to don and doff button-down shirts, pants, socks,

and shoes. He occasionally needed assistance to snap his pants but was able to zip, unzip, and unsnap. He was able to unscrew the toothpaste cap, but required assistance with resistive containers. In school H.N. used a built-up pencil holder and required extra time to complete tasks.

H.N. is now 17 years old and a senior in high school. He is currently in remission and off all medications except for Advil as needed. He did require bilateral hip replacements at age 15. Although his past years of severe active arthritis have left him with marked residual problems, he walks much better with the hip replacements and exhibits only a slight limp. His range and strength in both hands continue to be within functional limits. The strength in both the left and right hands reveals the following: lateral pinch equals 20 pounds, palmar pinch equals 17 pounds, and grip strength equals 65 pounds. He continues to posture his thumbs in slight MCP flexion; however, he exhibits full passive range of motion (Figure 9–48). H.N. states that he does not experience pain in his hands or arms, even

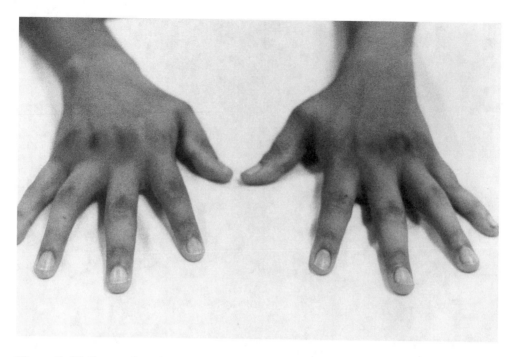

Figure 9–48 Range of motion of hands at 17 years of age.

with extreme weather changes. H.N.'s family has given exceptional support, encouraged his compliance with treatment regimens, and made efforts to understand JRA. H.N. has also been very cooperative throughout the years with both his medical and OT programs. This has resulted in his ongoing level of total independence and determination to complete high school and attend college.

REFERENCES

1. Scherzer A, Tscharnuter I. *Early Diagnosis and Therapy in Cerebral Palsy*. New York: Marcel Dekker; 1990.
2. Boehme R. *Improving Upper Body Control*. Tucson, AZ: Therapy Skill Builders; 1988.
3. Yasukawa A. Case report—upper extremity casting: adjunct treatment for a child with cerebral palsy hemiplegia. *Am J Occup Ther*. 1988;44:840–846.
4. Law M, et al. Neurodevelopmental therapy and upper extremity inhibitive casting for children with cerebral palsy. *Dev Med Child Neurol*. 1991;33:379–387.
5. Tona JL, Schneck CM. The efficacy of upper extremity inhibitive casting: a single-subject pilot study. *Am J Occup Ther*. 1993;47:901–910.
6. Tardieu C, et al. Muscle hypoextensibility in children with cerebral palsy, I: clinical and experimental observations. *Arch Phys Med Rehabil*. 1982;63:97–102.
7. Tardieu G, et al. Muscle hypoextensibility in children with cerebral palsy, II: therapeutic implications. *Arch Phys Med Rehabil*. 1982;63:103–107.
8. Cruickshank DA, O'Neill DL. Upper extremity inhibitive casting in a boy with spastic quadriplegia. *Am J Occup Ther*. 1990;44:552–555.
9. Smith LH, Harris SR. Upper extremity inhibitive casting for a child with cerebral palsy. *Phys Occup Ther*. 1985;5:71–79.
10. Gentile AM. Skill acquisition: action, movement, and neuromotor processes. In: Carr JH, Shepherd RB, eds. *Movement Science Foundations for Physical Therapy in Rehabilitation*. Gaithersburg, MD: Aspen Publishers; 1987.

11. Freehafer AA. Flexion and supination deformities of the elbow in tetraplegia. *Paraplegia.* 1977;15:221–225.

12. Moberg E. Surgical treatment for absent single hand grip and elbow extension in quadriplegia. *J Bone Joint Surg Am.* 1975;57:196–206.

13. Moberg E. The present state of surgical rehabilitation of the upper limb in tetraplegia. *Paraplegia.* 1987;25:351–356.

14. Abston S. Scar reaction after thermal injury and prevention of scars and contractures. In: Boswick JA, ed. *The Art and Science of Burn Care.* Gaithersburg, MD: Aspen Publishers; 1987: 359–371.

15. Bennett GB, Helm P, Purdue GF, Hunt JL. Serial casting: a method for treating burn contractures. *J Burn Care Rehabil.* 1989;10:543–545.

16. Bell-Krotoski JA, Figarola JH. Biomechanics of soft-tissue growth and remodeling with plaster casting. *J Hand Ther.* April–June 1995:131–137.

17. Melvin JL. *Rheumatic Disease in the Adult and Child: Occupational Therapy and Rehabilitation.* Philadelphia: F.A. Davis; 1989.

18. Emery H, Kucinski J. *Management of Juvenile Rheumatoid Arthritis: A Handbook for Occupational and Physical Therapists.* Chicago: LaRabida Children's Hospital and Research Center; 1987.

19. Chaplin D, Pulkki T, Saarimaa A, et al. Wrist and finger deformities in JRA. *Acta Rheumatol Scand.* 1969;15:206–223.

20. Findley TW, Halpern D, Easton J. Wrist subluxation in JRA: pathophysiology and management. *Arch Phys Med Rehabil.* 1983;64:69–73.

21. Athreya B. Hand in juvenile rheumatoid arthritis. *Arthritis Rheum.* 1977;20:573–574.

22. Nalebuff EA. The rheumatoid thumb. *Clin Rheum Dis.* 1984;10:589–608.

23. Bell JA. Plaster casting for the remodeling of soft tissue, part II. *The Star.* 1985;44(5):10–16.

CHAPTER 10

Medical Management of Spasticity

Puliyodil A. Philip and Mersamma Philip

Spasticity is defined as a motor disorder characterized by a velocity-dependent increase in tonic stretch reflexes (muscle tone) with exaggerated tendon jerks resulting from hyperexcitability of the stretch reflex as one component of the upper motor neuron (UMN) syndrome.[1] Muscle tone may be characterized as the sensation of resistance as one manipulates a joint through a range of motion, with the subject attempting to relax.[2] In a patient with spasticity, slowly applied stretch of a muscle may elicit little resistance, but as the speed of the stretch is progressively increased, resistance to the stretch progressively increases in magnitude. Spasticity may present either benefits or disadvantages (Exhibit 10–1).

PATHOPHYSIOLOGY OF SPASTICITY

The pathophysiology of spasticity is not well understood. Supraspinal and spinal pathways contribute to the increase in reflex excitability.[3–5] Dysfunctions of UMN pathways occur in conditions that involve insult to the cerebrum, brain stem, or spinal cord due to vascular, traumatic, infectious, neoplastic, or other means. Common etiologies include stroke syndromes, spinal cord injury, traumatic brain injury, multiple sclerosis, cerebral palsy, transverse myelitis, primary or metastatic central nervous system (CNS) tumors, or any lesion producing insult to the CNS. Initially, it was thought that the increase in stretch reflexes in spasticity resulted from hyperactivity of gamma motor neurons.

Recent studies have cast doubts on this explanation.[6] In some cases gamma overactivity may be present, but it is probable that changes in the direct input to alpha motor neurons and interneurons play a more important role. Thus the presence of spasticity is clear evidence of disordered descending input to motor neurons.[7] Modulation of transmission between the UMN and the lower motor neuron may result in reduction in the inhibitory effect of higher centers upon the motor unit. In spastic-

---

Exhibit 10–1 Effects of Spasticity

Disadvantages
  Alteration of balance
  Pain
  Joint contracture, subluxation, dislocation
  Impaired hygiene
  Pressure ulcers
  Decrease or loss of activities of daily living skills
  Interference with surgical healing
  Difficulty in positioning
  Weakness
Advantages
  Maintaining muscle bulk
  Decreased stasis edema
  Providing stability for standing and walking
  Reduced osteopenia
  Bowel training

ity there is an enhanced reflex response to muscle stretch. The segmental reflex arc consists of muscle receptors, their central connections with the spinal cord, and motor neuron output to muscle. Within this arc, the alpha motor neuron is the final conduit for motor neuronal outflow. This outflow is modulated by excitatory postsynaptic potentials from group IA and group II muscle spindle afferents, inhibitory postsynaptic potentials from interneuronal connections from the antagonist muscles, and presynaptic inhibition initiated by descending fiber output.[8]

Skeletal muscle consists of a large number of extrafusal muscle fibers under the control of gamma motor neurons. The length of the muscle is measured by specialized sensory nerve endings known as annulospiral endings of the muscle spindle apparatus. When the muscle spindle apparatus is elongated or stretched, a nerve impulse is generated through IA afferent fiber. This elicits an excitatory potential at the alpha motor neuron, resulting in contraction of extrafusal muscle fibers. This elongation–contraction interplay is the basis of the stretch reflex tested clinically. The annulospiral ending has nuclear bag and nuclear chain fibers. Nuclear bag fibers respond to dynamic changes in muscle length while nuclear chain fibers measure total muscle length. The sensitivity of the muscle spindle apparatus is modulated by the rate of firing of the intrafusal muscle fibers located in the terminal portion of the muscle spindle. Alpha stimulation causes the intrafusal fibers to contract, stretching the annulospiral endings and facilitating an IA afferent discharge. Sensory receptors located in the tendon known as Golgi tendon organs are autogenous IB inhibition receptors. As the tendon is stretched by the contracting muscle, IB afferents triggered by the Golgi tendon organs activate interneurons and inhibit motor neuron activity.[9,10] This is a complex interrelationship of motor systems with excitatory and inhibitory feedback loops. The knowledge of this system allows the clinician to perform appropriate evaluation and management for the individual with spasticity.

ASSESSMENT OF SPASTICITY

The goals of spasticity assessment are to identify the appropriate patients for specific types of therapy, to es-

tablish treatments, and to monitor changes in disability and impairments. A proper assessment of spasticity should include assessment of impairments, disability, and activities of daily living (ADLs). Tone is usually evaluated with the Ashworth or modified Ashworth scale[11,12] (Exhibit 10–2). Force production can be measured by manual muscle testing. The use of a hand-held force transducer or the amplitude of surface electromyography (EMG) to compare normal to abnormal states will increase sensitivity to the muscle contraction. Latency of activation, reciprocal inhibition, and inability to turn the muscle off are measured by surface EMG.

Loss of flexibility is assessed using range of motion measurements with a goniometer. Pain is measured using a scale such as the Borg Scale, where 0 is equivalent to no pain and 10 is equivalent to unbearable pain. Gait can be assessed using clinical evaluation or gait analysis in a gait lab. Other evaluations include assessment of disability and ADL functions.

MANAGEMENT OF SPASTICITY

The goal is to minimize the adverse effects of spasticity without compromising function.[13] All manifestations of spasticity can rarely be eliminated, and some spasticity is often beneficial. A stepped logarithmic approach for management of spasticity begins with conservative methods that carry few side effects, whereas

Exhibit 10–2 Definition of Ashworth Scores

Ashworth Score	Degree of Muscle Tone
1	No increase in tone
2	Slight increase in tone, resulting in a "catch" when affected limb is moved in flexion and extension
3	More marked increase in tone; passive movements difficult
4	Considerable increase in tone; passive movements difficult
5	Affected part rigid in flexion and extension

aggressive treatments have the most side effects.[14] A thorough clinical evaluation should reveal whether sufficient indications exist for one or more methods of management. In considering indications for treatment, one must simultaneously regard what useful effects of spasticity will be interrupted by its treatment. The side effects of all forms of management must be carefully considered, since they may negate these beneficial effects. The various medical management techniques and physical modalities, including casting, are often used in concert to maximize results. Some of the techniques discussed, such as rizotomy, surgery, pharmacological interventions, and biofeedback, focus on spasticity management while others (orthopaedic surgery and stretching) are focused on managing contracture.

Patient and family education should be provided illustrating the benefits and adverse effects of spasticity. Patients should also be instructed about how the spasticity can be used for functional activities such as transfers. Any treatable or preventable sources of nociception should be eliminated. These include pressure ulcers, ingrowing toenail, acute abdomen, bowel impaction, urinary tract infection, and any other sources that may increase spasticity.[14,15]

Physical Modalities

A daily stretching program prevents joint capsule tightness and contracture and helps restore resting length of muscle, tendon, and joint capsule.[16,17] The effects of a stretching program can often last for several hours. Maintaining the length using static stretching of muscles with splints, serial casting, or orthotics may be used.[16-20] In some instances, cutaneous stimulation from the orthosis may facilitate motoneuron activity, providing static stretching.[21] In many instances, orthoses may improve gait without sufficient direct reduction of spasticity.[22,23]

A number of neuromuscular facilitation techniques have been employed to provide motor reeducation, to enhance motor control, and to improve coordination.

Some of these techniques include those described by Bobarth and Bobath,[24,25] Rood,[26] and Kabat.[27] Various body or head positions can be used to minimize facilitation that is contributing to spasticity and maximize facilitation to muscles that have decreased voluntary recruitment.[28,29]

Vibration of large muscles has been shown to reduce spasticity. Tendon vibration may facilitate agonist voluntary movement and inhibit antagonist spasticity.[30-32]

Computed biofeedback provides a sensory modality feedback mechanism for patients to isolate and coordinate underlying agonistic and antagonistic musculature.[33,34] In theory, afferent stimulation is blocked by posterior column stimulation.[30]

Technical advances allow for exercise of spastic muscle groups and functional activities, including gait, using functional electrical stimulation of the skeletal muscles.[35,36] Electrical stimulation at nearly all levels of the nervous system relieves spasticity.[37] Stimulation of muscle or nerve for 15 minutes reduces spasticity and clonus for many hours, but functional gains have not been demonstrated.[38,39] Muscular reeducation using a combination of functional electrical stimulation and EMG-integrated biofeedback loops has been shown to improve strength and function in spastic limbs to a greater degree than when either modality is applied independently.[40]

Therapeutic heat and cold have been used in the management of spasticity. Cold applied to skin theoretically reduces the sensitivity of cutaneous mechanoreceptors that influence interneuron excitatory presynaptic potentials at the spinal cord level.[41] Cryotherapy is reported to reduce deep tendon reflexes and clonus.[42] Ultrasonic energy is used to preheat contracted soft tissue, including tendon, ligament, and joint capsule, prior to applying stretch.

Serial casting is used to manage abnormal tone and secondary contractures in the extremities. This text is dedicated to detailed description of this method of management.

Pharmacological Intervention

Pharmacological agents are an important component in the management of spasticity. No medication has been uniformly useful in the treatment of spasticity. All medications used to treat spasticity have potential toxicities with varying degrees of adverse effects in different patient populations. The cognitive and behavioral effects of these medications can be more problematic in patients with brain injuries than in those with spinal cord injuries. Continued use of pharmacologic agents should be contingent on a clearly beneficial effect. The site of action may make certain classes of antispasticity agents preferred for some conditions.[43,44] Pharmacological management produces the best effect when used with other conservative measures. The most commonly used medications are diazepam, baclofen, and dantrolene sodium.[45] Other drugs, somewhat less commonly used but becoming more popular, are clonidine and tizanidine. These medications are often in use while cast intervention is pursued. They have a more central, general effect on abnormal tone throughout the body while casting focuses on individual joints.

Dantrolene Sodium

Dantrolene sodium exerts its effect directly on skeletal muscle fibers, in contrast to centrally acting drugs such as diazepam and baclofen.[46] Because of this peripheral effect, dantrolene is the preferred agent for spasticity after brain injury, cerebral palsy, stroke, and other cerebral causes.[47] It acts peripherally on excitation-contraction coupling of muscle fibers.[48] Dantrolene reduces muscle tone by interfering with release of calcium from sarcoplasmic reticulum by reducing the excitation-contraction coupling necessary for muscle contraction.[49] Tendon jerk responses and electrically induced twitch tensions of muscle are reduced much more effectively than tetanic stimulation or sustained volitional contraction.[50] Of the most commonly used antispasticity medications (diazepam, baclofen, and dantrolene sodium), dantrolene is the least likely to cause severe lethargy or sedation, even though it may produce mild sedation during the initial treatment period. Dantrolene may be preferred for spasticity due to cerebral origin with a range of daily dosages. Because of hepatotoxicity, liver function tests should be monitored. Hepatic injury was noted in nearly 2% of patients, and death due to liver failure was seen in 0.3% of patients monitored for 60 days or more.[51,52] Adverse reactions are dependent on daily usage, duration of treatment, age of the patient, and condition of the liver. Dantrolene may be a useful adjunct for the treatment of spasticity after spinal cord injury.[53]

Baclofen

Baclofen appears to act on inhibitory synapses in the spinal cord that have γ-aminobutyric acid (GABA) as

the neurotransmitter, binding to $GABA_B$ receptors.[54] Baclofen is an inhibitor of the excitatory presynaptic potential, reducing stimulation of the anterior horn cell by the interneuron pool.[55] Baclofen inhibits both mono- and polysynaptic reflexes and also reduces activity of the gamma efferent.[56] The net result of the action of baclofen is to inhibit the firing pattern of the alpha motoneuron pool in the spinal cord with subsequent reduction of spasticity in skeletal muscles. Recent work has also demonstrated that part of its relaxant effect may be mediated through substantia nigra.[57] Baclofen crosses the blood-brain barrier, in contrast to GABA.[54,58]

Baclofen is probably the drug of choice in spinal forms of spasticity due to spinal cord injury and multiple sclerosis.[59,60] Baclofen seems to be particularly effective for the flexor and extension spasms commonly observed with spinal cord lesions.[45] Baclofen may be used as an adjunct in the treatment of spasticity after head injury, although this role remains unresolved.[56,61,62] Patients with acquired brain injury and elderly patients appear more likely to experience significant cognitive side effects. Baclofen decreases contraction of the external urethral sphincter and may improve bladder control.[63]

The approved daily oral dose ranges somewhat. Side effects include confusion, hallucinations, sedation, ataxia, and hypotonia.[61,64] Sudden withdrawal of the drug may lead to seizures.[65] The incidence of adverse effects increases with higher doses and older patients.[66]

Intrathecal administration of baclofen produces higher therapeutic concentrations of baclofen at potential sites of action within the spinal cord without incurring the concomitant systemic toxicity generally observed with oral administration.[67] A drug pump is implanted subcutaneously to infuse baclofen into the lumbar subarachnoid space. The pump with the reservoir is connected to an intrathecal catheter and the reservoir can be filled by percutaneous puncture. The rate of baclofen infusion can be adjusted by an external computer. In a study by Azouvi and associates, patients were followed for an average of 37 months.[68] A significant decrease in tone and spasms was observed in all patients. Functional assessment showed a highly significant ($P < 0.001$) increase of functional independence measure (FIM) score, particularly for bathing, dressing lower body, transfers, and, in some cases, locomotion. The efficacy remained stable after 6 to 9 months. Immediate and long-term benefits in the management of spasticity can be obtained with intrathecal baclofen use. Some patients experience complete disappearance of their spastic symptoms without any oral treatment.[69] In a study by Becker et al.,[70] patients who used intrathecal baclofen showed a decrease in the time spent in the acute care hospital. It was also found that the quality of life improved in areas such as transfer, seating, pain control, personal care, and decreased incidence of pressure ulcers.[70] Cervical intrathecal infusion of baclofen has also been described.[71] In this study, eight patients with head injury and three patients with cerebral palsy received cervical intrathecal infusion of baclofen. They showed improvement in mental status, speech, and dystonic and abnormal movements, and a reduction in spasticity. Significant decreases in upper and lower limb spasticity can be achieved by using continuous intrathecal baclofen infusion for spasticity of cerebral origin.[72] Some patients required significant increases in baclofen dosage to control spasticity during the first 12 months after implantation, but usually after that time no significant changes in dosage were required.[73,74] The therapeutic dosage requirements for patients with supraspinal spasticity are higher than for those with spinal spasticity.[75] Occasionally serious complications such as overdose, meningitis, and pump and tubing failure can occur.[76] Continuous intrathecal baclofen infusion used to manage severe spasticity in patients following hypoxic brain injury and in children with spasticity of cerebral origin showed patient improvement in spasticity as measured by the Ashworth Scale and functional mobility.[77,78]

Diazepam

Diazepam reduces spasticity by enhancing inhibitory effects of the central neurotransmitter, GABA, within the brain and spinal cord, perhaps at multiple levels of the CNS.[79,80] It reduces clonus, hypertonus, and flexor spasms. CNS side effects such as sedation and memory impairment are more pronounced with diazepam than with baclofen.[81] These limiting CNS effects often are most troublesome in head injury, stroke, encephalopathy, or multiple sclerosis.[82] Accommodations to these effects occur and many spinal cord–injured patients are treated successfully with higher doses.

Clonidine

Clonidine, a centrally acting α_2-receptor adrenergic agonist, has been successfully used as adjunctive therapy in patients with spinal cord injury who have problematic spasticity not adequately controlled by other antispastic agents.[83] Significant improvement in spastic hypertonia was observed, and transdermally delivered clonidine was well tolerated. Clonidine has also been used to treat spasticity due to brain stem lesions.[84] The adverse effects of clonidine are less with transdermal application, compared with oral use.[85] Gastrointestinal distress such as nausea, vomiting, and hypotension are the most common side effects of clonidine. Clonidine and baclofen seem to produce the best results in reducing spasticity in patients with spinal lesions in which management with a single medication does not give sufficient effect in reducing spasticity.

Tizanidine

Tizanidine is a centrally acting α_2-agonist, an imidazoline derivative structurally similar to clonidine. The antispasmodic activity of tizanidine is thought to result from indirect depression of polysynaptic reflexes, probably through the facilitation of the action of glycine, an inhibitory neurotransmitter antagonizing excitatory actions of the spinal interneurons.[86] It enhances vibratory inhibition of H-reflex and reduces abnormal cocontraction. The greatest reduction of spasticity coincides with peak serum concentrations. Tizanidine has been shown to be equivalent to baclofen as an antispastic agent, but it is better tolerated than diazepam in patients with chronic spasticity.[87-89]

Comparisons of tizanidine with active controls showed no differences in efficacy compared with baclofen or diazepam. However, when compared with controls, patients treated with tizanidine did not experience increased muscle weakness.[90,91] The most common adverse effects reported were dry mouth, somnolence, asthma, and dizziness. Mild elevations in liver function tests were also noticed occasionally in some patients, but these improved with dose reduction or withdrawal in all patients.

Other medications have been reported to have favorable antispasticity effects, and some have potential for treatment of patients with spasticity. Such drugs include ketazolam, threonine, orphenadrine citrate, mexiletine, chlorpromazine, and anticonvulsants such as phenytoin and gabapentin.

Nerve Blocks

Blocks can be classified as diagnostic or therapeutic. Diagnostic blocks are performed with a short-acting local anesthetic such as lidocaine or bupivacaine, allowing the examiner to evaluate passive range of motion, muscle stiffness, active range of motion, and motor control before and after the injection. If the patient shows reduction in spasticity, increase in range of motion, and improvement in function, then the decision can be made to proceed with a therapeutic block. These blocks are beneficial in affecting focal muscle groups and groups of muscles innervated by a given nerve. These blocks can be applied to a nerve trunk or its terminal nerve fibers (motor point block). A 2% to 6% aqueous phenol solution is used to produce chemical neurolysis.[92–96] Nerve blocks may be more effective and may last 3 to 6 months or more while neurolysis of motor points alone is generally less effective, perhaps due to the multiplicity of motor points within a muscle.[92,94,97,98] When percutaneous phenol blocks are done on pure motor nerves such as obturator or musculocutaneous nerve, complicating dysesthesias are not a problem. When mixed motor and sensory nerves receive injections at the level of plexus, root, or peripheral nerve, painful persistent dysesthesias may occur. Then intramuscular neurolysis (motor point block) or open phenol nerve block to isolate surgically only the motor branch of the mixed nerve is indicated.[93,98] Diluted phenol denatures protein and will destroy a portion of axons in any nerve bundle with which it comes in contact. The advantage of intramuscular and peripheral nerve neurolysis is that their effects are temporary, and repeated procedures may be done while the patient is recovering.[99] Casting is frequently used in conjunction with nerve blocks to manage underlying contracture once the block has acted to reduce the spasticity. Casts are usually applied 1 to 24 hours after the therapeutic nerve block in close consultation with the physician administering the block.

Botulinum Toxin Injection

Botulinum toxin type A, produced by *Clostridium botulinum*, causes blockade of the cholinergic junction at the motor end plate by inhibiting release of acetylcholine, functionally denervating the muscle.[100] The pharmacological properties of the botulinum toxin type A (BTX) depend primarily on the high-affinity binding of the type A toxin to human neuromuscular junction.[101] Return of normal muscle contraction occurs with regeneration and collateral sprouting of the nerve endings. The chemodenervation effect persists for many months. Clinical benefit generally lasts for 2 to 6 months, depending on the size and function of the muscle injected. The onset of clinical effect is typically seen 72 hours or more after an injection. Changes in tone, range of motion, and functional activities as measured by Barthel scores have been reported.[102] BTX has been reported as being useful in the management of not only spasticity but also muscle spasms and rigidity.[103] BTX has been shown to be effective in reducing spasticity and improving function in patients with stroke, multiple sclerosis, cerebral palsy, and other CNS disorders.[104–106] Casting is also frequently used in conjunction with BTX management to elongate the injected muscle and promote shortening of the antagonistic muscles. Casts are usually applied 10 to 14 days following BTX injection to ensure that maximal clinical effect has occurred.

Selective Posterior Rhizotomy

Posterior rhizotomy disrupts the spinal reflex arc but will not abolish spasticity mediated by suprasegmental pathways. Sectioning posterior roots is easier to perform than sectioning the anterior roots and does not result in denervation and muscle atrophy. Selective posterior rhizotomy (SPR) has been used primarily in patients with cerebral palsy.[107,108] SPR is advocated for children with purely spastic cerebral palsy who have reasonable voluntary motor control and trunk balance with no significant contractures and who are able to participate in an intensive therapy program after surgery; also for children who are nonambulatory and severely hypertonic in whom relief of spasticity may help positioning and daily personal care.[108,109] The disadvantages of SPR are associated ataxia, recurrence of spasticity, and sensory loss. A lumbar laminectomy is performed and dorsal rootlets are stimulated electrically; the decision to section is based on clinical observation or reflex muscle contraction. Tone is decreased in the majority of patients immediately following the surgery. Further therapy to upgrade mobility and other functions is required for longer periods of time. Bladder and bowel dysfunctions are avoided by sparing S2-S4 dorsal nerve roots. Physical therapy enhances functional improvement following SPR procedures in children with spastic diplegic cerebral palsy.[110] The complications with SPR include bronchospasm, aspiration pneumonia, urinary retention, ileus, sensory loss, and dysesthesia, and long-range complications such as decreasing range of muscle movement.[111,112]

Decreased postoperative migration of the hip in patients with spastic diplegic cerebral palsy has been reported following SPR.[113] Appropriate splinting, casting, and other therapeutic interventions will be necessary following SPR to upgrade the functions.

Other neurosurgical procedures that can be performed in extreme cases of spasticity include cordotomy, cordectomy, and cauda equina transection.[114] When useful voluntary motor control is preserved and when motor recovery is possible, such procedures are not appropriate. No evidence for reduction in spasticity for functional gains has been reported for either epidural or cerebellar stimulation.[115,116]

Orthopaedic Procedures

The goals of surgical treatment are to improve function and hygiene, facilitate total care of the patient, and provide pain relief. Patients with spasticity may have joint contractures, tendon and muscle contractures, joint subluxation or dislocation, and pain. Biceps lengthening, flexor-pronator release, and musculocutaneous neurectomy may improve elbow extension. Pectoralis major tenotomies and subscapularis lengthenings provide relief of shoulder pain and improved range of motion.[117] Spasticity of the wrist and hand causes weak grip, poor release of fingers, fixed thumb-in-palm deformity, and skin breakdown. Grip strength may be improved by repositioning the flexed wrist using tendon transfers, proximal row carpectomy, or radiocarpal arthrodesis. Lengthening of forearm flexor tendons may improve release of fingers after gripping.[118,119]

The most common deformity in the lower extremity related to spasticity is talipes equinovarus. The procedures used at the ankle and foot level include anterior

tibialis tendon–flexor hallucis longus tendon transfers, Achilles tendon lengthening, split anterior tibialis transfer, other tendon transfers and tendon releases, and triple arthrodesis.[120–123] Flexor contractures of the knee can be treated surgically by hamstring releases, rectus femoris release from the patella, and medial hamstring transfers.[118] Release of hip flexors may decrease muscle strength. If there is scissoring gait, it can be improved by psoas and adductor longus release and medial hamstring transfer.[124] Orthopaedic surgical correction of deformities and contractures serves no purpose if spasticity persists. As with casting to manage contracture, adjunct medical and other management to reduce spasticity is extremely important. While surgical release can be effective in correcting contractures and deformities, it is obviously invasive and generally utilized only after more conservative methods, including casting, have failed to result in goal achievement.

CONCLUSION

An understanding of the basic pathophysiology of spasticity is essential to determine appropriate treatment options. Reasonable treatment goals and long-term treatment strategies need to be discussed with the patient and family to achieve rehabilitation of the total patient. Many physical, pharmacological, and surgical interventions are available for treating the various manifestations of spasticity. Casting is one method. It is a method of intermediate conservativeness, more invasive than range of motion but less than nerve blocks and surgery. The selection of the appropriate treatment modality is dependent on the location and severity of spasticity, the goals of the patient, and the patient's motivation to be involved in the rehabilitation to achieve the goals. The patient's cognitive, emotional, and behavioral strengths and weaknesses also play a role in selecting the treatment strategies. Spasticity is best evaluated by studying functional movements and tasks. Technology-based assessments assist the clinician in sorting out many of these factors that affect individual functioning; however, the selection of the most appropriate methods for management of spasticity remains the challenge of the clinician. Many of the medical management strategies discussed are used in conjunction with casting to manage spasticity and associated contracture effectively.

REFERENCES

1. Lance JW. Symposium synopsis. In: Feldman RG, Young RR, Koella WP, eds. *Spasticity: Disordered Motor Control.* Chicago: Year Book Publishers; 1980:485–494.
2. Lance JW, McLeod JG. Disordered muscle tone. In: *Physiological Approach to Clinical Neurology.* Boston: Butterworths; 1981.
3. Burke D. Spasticity as an adaptation to pyramidal tract injury. *Adv Neurol.* 1988;47:401–423.
4. Young RR, Wiegner AW. Spasticity. *Clin Orthop.* 1987;219:51–62.
5. Katz R, Rymer WZ. Spastic hypertonia: mechanisms and measurement. *Arch Phys Med Rehabil.* 1989;70:144–155.
6. Burke D. A reassessment of the muscle spindle contribution to muscle tone in normal and spastic man. In: Feldman RG, Young RR, Koella WP, eds. *Spasticity: Disordered Motor Control.* Miami, FL: Symposia Specialists; 1980:261–278.
7. Gordon J. Receptors in muscle and their role in motor control. In: *The Physiological Basis of Rehabilitation Medicine.* 2nd ed. Stoneham, MA: Butterworth–Heinemann; 1994:103–125.
8. Katz RT. Mechanisms, measurements and management of spastic hypertonia after head injury. In: *Traumatic Brain Injury: Physical Medicine and Rehabilitation Clinics of North America.* Philadelphia: W.B. Saunders; 1992:319–335.
9. Brodal A. Spasticity: anatomical aspects. *Acta Neurol Scand.* 1962;38:8–40.
10. Parziale JR, Akelman E, Herz DA. Spasticity: pathophysiology and management. *Orthopedics.* 1993;16:801–811.
11. Ashworth B. Preliminary trial of carisoprodol in multiple sclerosis. *Practitioner.* 1964;192:540–542.
12. Bohannon RW, Smith MB. Interrator reliability of a modified Ashworth scale of muscle spasticity. *Phys Ther.* 1987;67:206–207.
13. Little JW, Massagli TL. Spasticity and associated abnormalities of muscle tone. In: *Rehabilitation Medicine: Principles and Practice.* 2nd ed. Philadelphia: J.B. Lippincott; 1993:666–680.
14. Merritt JL. Management of spasticity in spinal cord injury. *Mayo Clin Proc.* 1981;56:614–622.
15. Davis R. Spasticity following spinal cord injury. *Clin Orthop.* 1975;112:66–75.
16. Odeen I. Reduction of muscular hypertonics by long term muscle stretch. *Scand J Rehabil Med.* 1981;13:93–99.
17. Carey JR. Manual stretch: effect on finger movement control and force control in stroke subjects with spastic extrinsic finger flexor muscles. *Arch Phys Med Rehabil.* 1990;71:888–894.
18. Law M, Cadman D, Rosenbaum P, Walter S, Russell D, De Mattaeo C. Neurodevelopmental therapy and upper extremity inhibitive casting for children with cerebral palsy. *Dev Med Child Neurol.* 1991;33:379–387.
19. Feldman PA. Upper extremity casting and splinting. In: Glenn MB, Whyte J, eds. The practical management of spasticity in children and adults. Philadelphia: Lea & Febiger; 1990.
20. Connie TA, Sullivan T, Mackie T, Goodman M. Effects of serial casting for the prevention of equinus in patients with acute head injury. *Arch Phys Med Rehabil.* 1990;71:310–312.
21. Snook JH. Spasticity reduction splint. *Am J Occup Ther.* 1979;33:648–651.
22. Hinderer K, Harris S, Purdy A. Effects of tone reducing versus standard plaster casts on gait improvement of children with cerebral palsy. *Dev Med Child Neurol.* 1988;30:370–377.
23. Mossberg KA, Linton KA, Friske K. Ankle-foot orthoses: effect on energy expenditure of gait in spastic diplegic children. *Arch Phys Med Rehabil.* 1990;71:490–494.
24. Bobarth K, Bobath B. Treatment of cerebral palsy based on analysis of patients' motor behavior. *Br J Phys Med.* 1952;15:107–117.
25. Bobarth K, Bobath B. Cerebral palsy. In: Pearson H, Williams CE, eds. *Physical Therapy Services in Developmental Disabilities.* Springfield, IL: Charles C Thomas; 1972:131–175.
26. Rood MS. Neuromuscular mechanisms utilized in the treatment of neuromuscular dysfunction. *Am J Occup Ther.* 1956;10:220–225.
27. Kabat H. Studies on neuromuscular dysfunction. *Arch Phys Med Rehabil.* 1952;33:521–533.
28. Stejskal L. Postural reflexes in man. *Am J Phys Med.* 1979;58:1–25.
29. Bobarth K. *A Neurophysiological Basis for the Treatment of Cerebral Palsy.* 2nd ed. Philadelphia: J.B. Lippincott; 1980.
30. Griffin J. Use of proprioceptive stimuli in therapeutic exercises. *Phys Ther.* 1974;54:1072–1078.
31. Hagbarth KE, Eklund G. The muscle vibrator: a useful tool in neurological therapeutic work. *Scand J Rehabil Med.* 1969;1:26–34.
32. Giebler KB. Physical modalities. In: Glenn MB, Whyte J, eds. *The Practical Management of Spasticity in Children and Adults.* Philadelphia: Lea & Febiger; 1990.
33. Cozean CD, Pease WS, Hubbell SL. Biofeedback and functional electrical stimulation in stroke rehabilitation. *Arch Phys Med Rehabil.* 1988;69:401–405.
34. Shumway-Cook A, Anson D, Haller S. Postural sway biofeedback: its effect on reestablishing stance stability in hemiplegic patients. *Arch Phys Med Rehabil.* 1988;69:395–400.
35. Peterson T, Klemar B. Electrical stimulation as a treatment of lower limb spasticity. *J Neurol Rehabil.* 1988;2:103–108.
36. Sipski ML, DeLisa JA, Scheer S. Functional electrical stimulation bicycle ergometry: patient perceptions. *Am J Rehabil Med.* 1989;68:147–149.
37. Stefanovska A, Rebersek S, Bajd T, Vodovnik L. Effects of electrical stimulation in spasticity. *Phys Rehabil Med.* 1991;3:59–99.
38. Vodovnik L, Bowman BR, Winchester P. Effect of electrical stimulation on spasticity in hemiparetic patients. *Int Rehabil Med.* 1984;6:153–156.
39. Walker JB. Modulation of spasticity: prolonged suppression of a spinal reflex by electrical stimulation. *Science.* 1982;216:203–204.
40. Field RW. Electromyographically triggered electrical muscle stimulation for chronic hemiplegia. *Arch Phys Med Rehabil.* 1987;68:407–414.

41. Bell KR, Lehmann JF. Effect of cooling on H and T reflexes in normal subjects. *Arch Phys Med Rehabil.* 1987;68:490–493.

42. Knutson E, Mattsson E. Effects of local cooling on monosynaptic reflexes in man. *Scand J Rehabil Med.* 1969;2:159–163.

43. Clemente C. Neurophysiologic mechanisms and neuroanatomic substrates related to spasticity. *Neurology.* 1978;28:40–44.

44. Davidoff RA. Pharmacology of spasticity. *Neurology.* 1978;28:44–51.

45. Whyte J, Robinson KM. Pharmacologic management. In: Glenn MB, Whyte J, eds. *The Practical Management of Spasticity in Children and Adults.* Philadelphia: Lea & Febiger; 1990.

46. Monster AW, Herman R, Meeks S. Co-operative study for assessing the effects of a pharmacologic agent on spasticity. *Am J Phys Med.* 1973;52:163–188.

47. Ketel WB, Kolb ME. Long term treatment with dantrolene sodium of stroke patients with spasticity limiting the return of function. *Curr Med Res Opin.* 1984;9:161–169.

48. Ward A, Chaffman MO, Sorkin EM. Dantrolene: a review of its pharmacodynamic and pharmacokinetic properties and therapeutic use in malignant hyperthermia, the neuroleptic malignant syndrome and an update of its use in muscle spasticity. *Drugs.* 1986;32:130–168.

49. Lietman PS, Haslam RHA, Walcher JR. Pharmacology of dantrolene sodium in children. *Arch Phys Med Rehabil.* 1974;55:388–392.

50. Hermann R, Mayer N, Mecomber SA. The pharmacology of dantroline sodium. *Am J Phys Med.* 1972;51:296–311.

51. Wilkinson SP, Portmann B, Williams R. Hepatitis from dantrolene sodium. *Gut.* 1979;20:33–36.

52. Chan CH. Dantrolene sodium and hepatic injury. *Neurology.* 1990;40:1427–1432.

53. Hermann R, D'Luzansky SC. Pharmacologic management of spinal spasticity. *J Neurol Rehabil.* 1991;5:15–20.

54. Koella WP. Baclofen: its general pharmacology and neuropharmacology. In: Feldman RG, Young RR, Koella WP, eds. *Spasticity: Disordered Motor Control.* Chicago: Year Book Medical Publishers; 1980:383–396.

55. Jones R, Lance J. Baclofen in the long term management of spasticity. *Med J Aust.* 1976;1:654–657.

56. Van Hemert JCJ. A double blind comparison of baclofen and placebo in patients with spasticity of cerebral origin. In: Feldman RG, Young RR, Koella WP, eds. *Spasticity: Disordered Motor Control.* Chicago: Year Book Publishers; 1980.

57. Tuvski L, Klockgether T, Schwartz M. Substantia nigra: a site of action of muscle relaxant drugs. *Ann Neurol.* 1990;28:341–348.

58. Faigle JW, Keberle H, Degen PH. Chemistry and pharmacokinetics of baclofen. In: Feldman RG, Young RR, Koella WP, eds: *Spasticity: Disordered Motor Control.* Chicago: Year Book Medical Publishers; 1980.

59. Duncan GW, Shahani BT, Young RR. An evaluation of baclofen treatment for certain symptoms in patients with spinal cord lesions. *Neurology.* 1976;26:441–446.

60. Sachais BA, Logue JN, Carly MS. Baclofen, a new antispastic drug. *Arch Neurol.* 1977;34:422–428.

61. Hattab JR. Review of European clinical trials with baclofen. In: Feldman RG, Young RR, Koella WP, eds. *Spasticity: Disordered Motor Control.* Chicago: Year Book Medical Publishers; 1980.

62. Knutsson E, Lindblom U, Martensson A. Liorsal and spasticity. *Acta Neurol Scand.* 1972;48:449–450.

63. Khanna OP. Neurosurgical therapeutic modalities. In: Krane RJ, Siroky MB, eds. *Clinical Neurology.* Boston: Little, Brown and Co.; 1979.

64. Roy CW, Wakefield IR. Baclofen pseudopsychosis: case report. *Paraplegia.* 1986;24:318–321.

65. Terrence DV, Fromm GH. Complications of baclofen withdrawal. *Arch Neurol.* 1981;38:588–589.

66. Aisen ML, Dietz MA, Rossi P, Cedarbaum JM, Kutt H. Clinical and pharmacokinetic aspects of high dose oral baclofen therapy. *J Am Paraplegia Soc.* 1993;15:211–216.

67. Penn RD. Intrathecal baclofen for severe spasticity. *Ann NY Acad Sci.* 1988;531:157–166.

68. Azouvi P, Mane M, Thiebaut JB, Denys P, Remy-Nevis O, Bussel B. Intrathecal baclofen administration in severe spinal spasticity: functional improvement and long term follow up. *Arch Phys Med Rehabil.* 1996;77:35–39.

69. Pirotte B, Heilporn A, Joffroy A, et al. Chronic intrathecal baclofen in severely disabling spasticity: selection, clinical assessment and long-term benefit. *Acta Neurologica Belg.* 1995;95:216–225.

70. Becker WJ, Harris CJ, Long ML, Ablett DP, Klein GM, DeForge DA. Long-term intrathecal baclofen therapy in patients with intractable spasticity. *Can J Neurol Sci.* 1995;22:208–217.

71. Concalves J, Garcia-March G, Sanchez-Ledesma MJ, Onzain I, Broseta J. Management of intractable spasticity of supraspinal origin by chronic cervical intrathecal infusion of baclofen. *Stereotact Funct Neurosurg.* 1994;62:108–112.

72. Albright AL, Barron WB, Fasick MP, Polinko P, Janosky J. Continuous intrathecal baclofen infusion for spasticity of cerebral origin. *JAMA.* 1993;270:2475–2477.

73. Akman MN, Loubser PG, Donovan WH, O'Neill ME, Rossi CD. Intrathecal baclofen: does tolerance occur? *Paraplegia.* 1993;31:516–520.

74. Lewis KS, Mueller WM. Intrathecal baclofen for severe spasticity secondary to spinal cord injury. *Ann Pharmacother.* 1993;27:767–774.

75. Saltueri L, Kronenberg M, Marosi MJ, et al. Long-term intrathecal baclofen treatment in supraspinal spasticity. *Acta Neurol.* 1992;14:195–207.

76. Teddy P, Jamons A, Gardner B, Wang D, Silver J. Complications of intrathecal baclofen delivery. *Br J Neurosurg.* 1992;6:115–118.

77. Becker R, Alberti O, Bauer BL. Continuous intrathecal baclofen infusion in severe spasticity after traumatic or hypoxic brain injury. *J Neurol.* 1997;244:160–166.

78. Armstrong RW, Steinbok P, Cochrane DD, Kube SD, Fife SE, Farrell K. Intrathecally administered baclofen for treatment of children with spasticity of cerebral origin. *J Neurol.* 1997;87:409–414.

79. Verrier M, Ashby P, MacLeod S. Effect of diazepam on muscle contraction in spasticity. *Am J Phys Med.* 1976;55:184–187.

80. Davidoff RA. Mode of action of antispasticity drugs. *Neurosurg Rev.* 1989;4:315–324.

81. Roussan M, Terrance C, Fromm G. Baclofen versus diazepam for the treatment of spasticity and long term follow up of baclofen therapy. *Pharmacotherapeutica.* 1985;4:278–284.

82. Glenn MB, Wroblewski B. Antispasticity medication in the patient with traumatic brain injury. *J Head Trauma Rehabil.* 1986;1:71–72.

83. Yablon SA, Sipski ML. Effect of transdermal clonidine on spinal spasticity: a case series. *Am J Phys Med Rehabil.* 1993; 72:154–157.

84. Sandford PR, Spengler SE, Sawasky KB. Clonidine in the treatment of brainstem spasticity: case report. *Am J Phys Med Rehabil.* 1992;71:301–303.

85. Weingarden SI, Belen JG. Clonidine transdermal system for treatment of spasticity in spinal cord injury. *Arch Phys Med Rehabil.* 1992;73:876–877.

86. Coward DM. Tizanidine: neuropharmacology and mechanism of action. *Neurology.* 1994;44:6–11.

87. Hennies OL. A new skeletal muscle relaxant (DS103–282) compared to diazepam in the treatment of muscle spasm of local origin. *J Int Med Res.* 1981;9:62–68.

88. Medici M, Pebet M, Ciblis D. A double-blind long term study of tizanadine in spasticity due to cerebrovascular lesions. *Curr Med Res Opin.* 1989;11:398–407.

89. Lataste X, Emre M, Davis C, Groves L. Comparative profile of tizamidine in the management of spasticity. *Neurology.* 1994; 44:53–59.

90. Wallace JD. Summary of combined clinical analyses of controlled clinical trials with tizanidine. *Neurology.* 1994;44:60–69.

91. Smith C, Birnbaum G, Carter JL, Greenstein J, Lublin FD. Tizanidine treatment of spasticity caused by multiple sclerosis: results of a double-blind, placebo-controlled trial: US Tizanidine Study Group. *Neurology.* 1994;44:34–43.

92. Easton JKM, Ozel T, Halpern D. Intramuscular neurolysis for spasticity in children. *Arch Phys Med Rehabil.* 1979;60:155–158.

93. Garland DE, Lilling M, Keenan MA. Percutaneous phenol blocks to motor points of spastic forearm muscles in head-injured adults. *Arch Phys Med Rehabil.* 1984;65:243–245.

94. Khalili AA, Betts HB. Peripheral nerve block with phenol in the management of spasticity. *JAMA.* 1967;200:1155–1157.

95. Khalili AA. Physiatric management of spasticity by phenol nerve and motor point block. In: Ruskin AP, ed. *Current Therapy in Physiatry.* Philadelphia: W.B. Saunders; 1984.

96. Petrillo CR, Chu DS, Davis SW. Phenol block of tibial nerve in the hemiplegic patient. *Orthopedics.* 1980;3:871–874.

97. McComas AJ, Keveshi S, Manzano G. Multiple innervation of human muscle fibers. *J Neurol Sci.* 1984;64:55–64.

98. Garland DE, Lucie RS, Waters RL. Current uses of open phenol nerve block for adult acquired spasticity. *Clin Orthop Relat Res.* 1982;165:217–222.

99. Bottle MJ, Abrams RA, Bodine-Fowler SC. Treatment of acquired muscle spasticity using phenol peripheral nerve blocks. *Orthopedics.* 1995;18:151–159.

100. Botulinum toxin for occular muscle disorders. *Med Lett.* 1990;32:100–102.

101. Coffield JA, Considine RV, Simpson LL. The site and mechanism of action of botulinum neurotoxin. In: Jankovic J, Hallet M, eds. *Therapy with Botulinum Toxin.* New York: Marcel Dekker; 1994:3–13.

102. Snow BJ. Treatment of spasticity with botulinum toxin: a double-blind study. *Ann Neurol.* 1990;28:512–515.

103. Gracko MA, Polo KB, Jabbari B. Botulinum toxin A for spasticity, muscle spasms and rigidity. *Neurology.* 1995;45:712–717.

104. Hesse S, Lucke D, Malezic M, et al. Botulinum toxin treatment for lower limb extensor spasticity in chronic hemiparetic patients. *J Neurol Neurosurg Psychiatry.* 1994; 57:1321–1324.

105. Borg-Stein J, Pine ZM, Miller JR, Brin MF. Botulinum toxin for treatment of spasticity in multiple sclerosis: new observations. *Am J Phys Med Rehabil.* 1993;72:364–368.

106. Koman LA, Mooney JF, Smith B, Mulvaney T. Management of cerebral palsy with botulinum A toxin: preliminary investigation. *J Pediatr Orthop.* 1973;13:489–495.

107. Peacock WJ, Arens LJ. Selective posterior rhizotomy for the relief of spasticity in cerebral palsy. *South Afr Med J.* 1982; 62:119–124.

108. McDonald CM. Selective dorsal rhizotomy: a critical review. *Phys Med Rehabil Clin North Am.* 1991;2:891–915.

109. Abbott R, Forem SL, Johann M. Selective posterior rhizotomy for the treatment of spasticity: a review. *Child Nerve Syst.* 1989;5:337–346.

110. Steinbok P, Reiner AM, Beauchamp R, Armstrong RW, Cochrane DD. A randomized clinical trial to compare selective posterior rhizotomy plus physiotherapy with physiotherapy alone in children with spastic diplegic cerebral palsy. *Dev Med Child Neurol.* 1997;39:178–184.

111. Abbott R. Complication with selective posterior rhizotomy. *Pediatr Neurosurg.* 1992;18:43–47.

112. Abbott R, Johann-Murphy M, Shimiuski-Maher T, et al. Selective posterior rhizotomy: outcome and complications in treating spastic cerebral palsy. *Neurosurg.* 1993; 33:851–857.

113. Park TS, Volger GP, Phillips LH, et al. Effects of selective dorsal rhizotomy for spastic diplegia on hip migration in cerebral palsy. *Pediatr Neurosurg.* 1994;20:43–49.

114. Smolik EA, Nash FP, Machek O. Spinal cordectomy in the management of spastic paraplegia. *Am Surg.* 1960;26:639–645.

115. Gottlieb GL, Myklebust BM, Stefoski D. Evaluation of cervical stimulation for chronic treatment of spasticity. *Neurology.* 1985;35:699–704.

116. Penn RD. Chronic cerebellar stimulation for cerebral palsy: a review. *Neurosurgery.* 1982;10:116–121.

117. Braun RM, West F, Mooney V, Nichol BL, Rooper B, Caldwell C. Surgical treatment for the painful shoulder contracture in the stroke patient. *J Bone Joint Surg.* 1971;52:1307–1312.

118. Braun RM, Vise GT, Roger B. Preliminary experience with superficialis to profundis tendon transfers in the hemiplegic upper extremity. *J Bone Joint Surg.* 1974;56:466–472.

119. Smith RJ. Flexor pollicis longus adductor-plasty for spastic thumb-in-palm deformity. *J Hand Surg.* 1982;7:327–334.

120. Pinzur MS, Sherman R, Demonte-Levine P, Kett N, Trimble J. Adult-onset hemiplegia: changes in gait after muscle balancing procedures to correct equinus deformity. *J Bone Joint Surg.* 1986;68:1249–1257.

121. Bleck EE. Management of lower extremities in children who have cerebral palsy. *J Bone Joint Surg.* 1990;72:140–144.

122. Scott SM, Janes PC, Stevens PM. Grice subtalar arthrodesis followed to skeletal maturity. *J Pediatr Orthop.* 1988;8:176–183.

123. Green NE, Griffin RP, Shiavi R. Split posterior tibial tendon transfer in spastic cerebral palsy. *J Bone Joint Surg.* 1983;65:748–754.

124. Waters RL, Botte MJ, Jordan C, Perry J, Pinzur MS. Rehabilitation of stroke patients—the role of the orthopedic surgeon. *Contemp Orthop.* 1990;20:311–348.

CHAPTER 11

Upper Extremity Surgery

Audrey M. Yasukawa

Orthopaedic surgery to improve motor control of the upper extremity is very complex, especially with individuals with central nervous system (CNS) lesions. CNS lesions alter the coordination and control to produce various arm and hand patterns and actions. The action of voluntarily contracting and relaxing a hand when it is influenced by abnormal muscle pull or spasticity is difficult. Often, certain muscles are weak or absent, thus producing a muscle imbalance that may further lead to a myostatic contracture.

Serial casting may have been attempted initially, with prescription of an orthosis and an exercise program for follow-up. Depending on the individual patient, such a treatment regimen may result in only minor and temporary improvement. Often the lack of any substantial improvement is recognized by the patient and may eventually lead to discontinued wearing of an orthotic device.

Surgical options should be considered when the benefits of a standard occupational therapy or rehabilitation program have reached their limits. Selection for surgical procedure is based on what the client, therapist, and physician believe will produce the best results for the individual's function. This decision is based on the physician's and therapist's observation of the client's hand use and the client's goals and expectations about surgery. The decision to pursue surgery is usually made after many years of therapy when the patient has developed definitive patterns of arm and hand movements.

ASSESSMENT

Evaluation by the physician and occupational therapist is critical in assessing motor skills and deficiency of function in the patient's arm. For a person influenced by spasticity, muscle imbalance, or compensatory patterns, it may be difficult to evaluate specific muscle control and phasic activity of the various muscles. A combination of muscles may be involved, and all or some of them may require surgical correction. The limb as a whole should be assessed. Problems of the hand should be addressed after evaluating the ability to move and orient the hand in space. The function and alignment of the shoulder, elbow, and forearm must be assessed in relation to the position of the hand. In addition, one must look at the wrist and fingers as they support the function of the thumb.

Several nonsurgical techniques can be used to assist the physician and therapist in determining the appropriateness of surgical procedures. Electromyography (EMG) can sometimes be used to assist in the selection for potential muscle tendon transfers.[1-4] A muscle or nerve block is another alternative to assist with determining the presence or absence of an underlying fixed contracture, to predict the outcome of planned muscle releases, and to be used in conjunction with splinting and casting.[5,6]

Other factors to assess prior to surgery include the patient's motivation to participate in a postoperative program, sensibility of the hand, and type and extent of the CNS lesion. More than one procedure may be done at the same time, or a sequence of surgeries may be needed to alter several muscles. Every case must be individualized according to the priorities of the patient and the family. The optimal candidate is well motivated, with sensory awareness of the hand as well as some voluntary control and use of the upper extremity.

SURGICAL INDICATIONS AND CONTRAINDICATIONS

The motivation of the patient is one of the most important prerequisites for successful surgical treatment. Other surgical indications may include (1) the patient's ability to participate actively in a rehabilitation program, (2) the patient's having sufficient cognitive level to follow-up with the program, (3) the function of a nonfunctional hand could be improved through a successful surgical procedure, (4) correction of the flexed posture of the wrist and fingers could improve either function or hygiene, (5) cosmetic appearance could be improved in a patient for whom an improved appearance would be of significant value, and (6) the patient's need to wear a splint permanently could be eliminated.

The contraindications for surgery may include (1) insufficient trial of a more conservative measure, (2) excessive joint instability of all joints of the hand, (3) severe athetosis, (4) age less than 5 years, (5) emotional problems, (6) low cognitive level and consequent inability to follow rehabilitation instructions, (7) hand function that would not be expected to improve by surgery, (8) severe sensory deficit (although this is not a contraindication for improving hygiene or appearance by surgery), (9) lack of voluntary control of the arm and hand muscles, and (10) stabilization of a joint that would interfere with the present functional control of the patient as he or she is using the hand for grasp and release (to position the wrist for assisting with a function), transferring, and/or ambulating with crutches or a walker.

SURGICAL OPTIONS

The various options and guidelines listed below are useful when examining a patient. They are, however, only general guidelines, and an extensive period of evaluation is required to determine an appropriate surgical treatment. It is not the purpose of this text to provide detailed surgical procedures and methods of management. Numerous articles and books are available to assist a therapist in understanding the complexity of surgical procedures, rationale and outcome results.[7–10]

Generally the surgical protocol starts with selective tenotomies to release the deforming force. Next is a rebalancing of the muscles, usually between the spastic flexors and weakened extensors, which may be accomplished by a muscle transfer. The physician and occupational therapist must evaluate the action of the weakened or paralyzed muscle with that of the muscle to be transferred. However, tendon transfer alone can never overcome rigid osseous deformity. In such cases the final resort may be an arthrodesis.

Various types of surgical options may be performed on the upper limb. The following review of commonly used surgical procedures and orthopaedic classification systems will assist therapists when they communicate with other medical professionals and may be a useful baseline for evaluating the patient's potential for function.

Flexion Contracture of the Elbow

An elbow flexion contracture greater than 45° may pose significant problems for functional reach. Candidates may benefit from Z-lengthening of the bicipital tendinous insertion, release of the lacertus fibrosus, flexor origin slide, an aponeurotic section, or myotomy.[11–13] Walters[6] describes a technique for longstanding elbow flexion deformities, greater than 75°. The origin of the brachioradialis is released, the bicep tendon is released, and a myotomy is performed on the brachialis muscle. During the surgery, less than 45° of correction is performed to prevent tension on the shortened median nerve or brachial artery. However, after surgery further correction can be obtained by serial casting.

Pronation Contracture of the Forearm

Pronation contracture places the client at a biomechanical disadvantage for functional grasp and release. The angle of the pronated forearm limits the ability to orient the hand to promote functional prehension. If the supinator muscles are weakened or paralyzed, it has been my experience with children with cerebral palsy that attempts to treat the fixed pronation deformities by immobilization in a long arm cast followed by an aggressive exercise program have been unsuccessful. However, for patients with spinal cord injury (SCI), long arm casts have been successful but long-term splint use is needed to maintain this position. Surgery for patients with SCI is helpful for active supination and maintenance without the use of a splint. Forearm stability is a very important function gained from the surgery.

A Green's transfer[14] in which the flexor carpi ulnaris (FCU) is transferred to the extensor carpi radialis brevis (ECRB) to improve wrist extension is believed to support forearm supination. This operation is used primarily to correct flexion deformity of the wrist and is thought to fail to correct the pronation contracture. Sakellarides et al.[15] discussed a surgical technique by changing the insertion of the pronator radii teres into an active forearm supinator.

Gschwind and Tonkin[16] introduced a classification system for evaluating patients with a pronator deformity with suggested treatment options. The groups are as follows:

I.   Patient can actively supinate beyond neutral. No specific surgery is indicated.
II. Patient can actively supinate to neutral or less than neutral. The patient may benefit from a release procedure of the pronator quadratus or flexor aponeurosis. Possibly any remaining imbalance may necessitate an active supinator transfer. However, care must be taken to avoid a supinator contracture.
III. Patient cannot actively supinate; however, full passive range of motion exists. There is no motor control present and a pronator teres transfer will assist the movement for supination.
IV. Patient cannot actively supinate and there is tightness in passive range of motion. There may be myostatic contractures and tightness in the interosseous membrane, joint malalignment, radial head subluxation, and muscle imbalance of the pronators and supinators. The patient may benefit from a release procedure of the pronator quadratus or flexor aponeurosis. Evaluate whether an active transfer is also needed.

Flexion Contractures of the Wrist and Fingers

Zancolli and Zancolli[17] have developed a classification system of the spastic hand of infantile cerebral palsy that is useful in determining the type of surgical procedure to be recommended. This classification groups deformities according to the patient's ability to extend the wrist and fingers. Tonkin and Gschwind[18] used Zancolli and Zancolli's classification system as a guideline on which to base surgical decisions. They reported the results of 34 patients treated for flexion deformities of the wrist and fingers. Of 34 patients, 30 were found to improve functionally and cosmetically.

Zancolli and Zancolli's classification system consists of the following:

1. The patient can extend the fingers with the wrist in neutral or with less than 20° of flexion. The patient is unable to extend the fingers with the wrist in dorsiflexion. Flexion spasticity is localized at the FCU muscle. The surgical procedure may consist of a tenotomy of the distal tendon of the FCU and an aponeurotic release of the origin of the medial epicondyle muscle.
2. The patient can extend the fingers with the wrist in greater than 20° of wrist flexion. Spasticity is localized at the wrist and finger flexors. This is further subdivided as follows:
    2a. The patient can dorsiflex the wrist with the fingers flexed. This implies that the wrist extensors are active; however, the spasticity is more localized at the finger flexors.
    2b. The patient demonstrates paralysis of the wrist extensors and cannot extend the wrist with the fingers in flexion. The surgical procedures may consist of an FCU tendon transfer to the ECRB and an aponeurotic release of the medial epicondyle muscle.
3. The patient demonstrates marked deformities of the arm into flexion and pronation. There are no active wrist or finger movements. Surgery for this patient is primarily indicated for hygiene or comfort. Surgical procedures may consist of flexor muscle origin release or lengthening the flexor tendons of the wrist and fingers. It may also be necessary to correct the pronation and elbow flexion contractures.

The procedure of Green and Banks[14] is a surgical technique used to improve wrist function. The patient must have active finger extensors for good results with this transfer. In general, transfer of the FCU to the extensor carpi radialis longus (ECRL) or ECRB is advisable when the wrist extensors are weak. McCue et al.[19] described the use of transferring the brachioradialis to the ECRL and ECRB, extensor pollicis longus (EPL), flexor digitorum profundus, and extensor digitorum communis (EDC), and for thumb opposition for hand deformities in 33 tendon transfers, in which 28 hands were improved.

If the basic problem in the hand is the inability to release secondary to poor finger extensors, a transfer to the EDC, rather than ECRB, should be helpful.[20,21]

Thumb Deformities

An intricate balance exists with the alignment and stability of the wrist and function of the thumb. The ability to use the thumb is essential for functional prehension.

House et al.[22] developed a classification system of the thumb and the various problems encountered with the treatment of thumb deformities. The system classification consists of the following:

Type I.   The metacarpal (MC) is held in adduction. This is the most common problem secondary to spasticity and fixed contracture of the first dorsal interosseus (DI) and adductor pollicis. Contracture of the web space may also exist. Surgical procedures may include a release of the adduction-flexion deformity, lengthening to allow greater abduction and extension of the thumb, Z-plasty of the skin of the thumb cleft, or release of the first DI.

Type II. The MC is held in adduction with flexion of the metacarpophalangeal (MCP) joint. There is spasm and contracture of the flexor pollicis brevis (FPB). Surgical procedures include release of the origin of the adductor pollicis and FPB, Z-plasty of the thumb cleft, and selective transfers of a tendon to augment a weak muscle. The muscle transfer to assist with extrinsic abduction-extension power will depend on the extent of control of the muscle. Generally the palmaris longus, brachioradialis, and flexor carpi radialis (FCR) are used. This will depend on which of the three joints in the thumb will need balancing by a muscle transfer force. The motor transfer used may be inserted into one or more of the tendons of the extensor pollicis brevis or longus, or the abductor pollicis longus.

Type III. The MC is held in adduction, the MCP is unstable, and hyperextension deformity is noted during active use. The surgical procedure may require stabilization of the joints if a functional balance cannot be attained by a tendon transfer.

Type IV. The MC is adducted and the MCP and the interphalangeal (IP) joints are tightly flexed into the palm. It is often the result of spasticity of the flexor pollicis longus (FPL) and contractures of the intrinsic muscles of the thumb. Surgical procedures may consist of lengthening the FPL, release of the adductor pollicis and FPB, Z-plasty in the cleft of the thumb. For a tight FPL and IP joint flexed into the palm, the EPL may not have an appropriate muscle strength to maintain the extended position even after lengthening of the FPL. Augmentation of the EPL may be required to assist with both abduction and extension of the distal joint.

After a careful screen, evaluation of motor strength, deficiency, and contractures, a pretrial cast may assist in the planning of the surgical treatment. Many patients have combined deformities and therefore careful consideration of all surgical options is important.

Casting Program

The use of an upper extremity cast prior to surgery may provide valuable information. Such a pretrial cast can assist with the selection of the most appropriate operative procedure and can help determine the probable results and emotional response of the patient.

Flexion deformity of the hand, the most commonly encountered problem, is caused by the predominance of spasm in the flexor muscles of the wrist, fingers, and thumb. Generally when the deformity has been present for a prolonged period of time, soft tissue contractures occur with shortening of the skin, ligaments, tendons, and muscle. Through a series of casts, the wrist can be stretched gradually to full passive range into wrist extension. A mobile wrist is essential to provide a functional grasp and release when the decision is to perform a tendon transfer to augment wrist extension. In the presence of contractures and tight wrist flexors, the alternative may be an arthrodesis. The surgical decision will be indicated by the results of the casting trial.

Prior to cast application, the therapist should note the degree of wrist flexion needed to extend the fingers completely or partially. Can the patient extend the wrist with the fingers flexed? Also note the degree of tightness of the wrist, fingers, or thumb. Will the patient need lengthening of the fingers or thumb in addition to having other procedures, such as tendon transfers? Oftentimes, the wrist and fingers will require surgery in addition to the thumb. The therapist observes the control of the fingers and thumb for functional grasp and release with the wrist casted in neutral or in slight dorsiflexion. While positioned in the cast the patient can strengthen and reeducate the finger flexors and extensors to promote a functional grasping mechanism.

Generally an arthrodesis of the wrist or thumb joint is performed as a last resort if a functional balance cannot be achieved by a transfer. A wrist cast with the thumb enclosed can assist in determining the ideal position for the arthrodesis of a joint.

Since the aim of a surgical procedure is to produce a good grasp and release and pinch, the extent of voluntary control while in the cast can indicate the potential for active use of the hand. The goal may be one of improving function of the involved extremity so that it can be used as a better stabilizer and helper for the more functional limb. If a client has absent to poor muscle control, surgical treatment may be directed toward promoting hygiene or cosmesis with use of the hand as a passive assist at best. Goals such as improved gross hand function to operate a communication board or improved appearance may be gratifying for the pa-

tient. The family's and client's understanding of the long-term goals of surgery are essential.

POSTOPERATIVE PROGRAM

A carefully planned follow-up program is crucial for the patient's postoperative success. A postoperative treatment program may range in duration depending on the type of surgical motor transfer or procedure and the functional outcome. The occupational therapist's role is to augment or improve the surgical results. Generally, 4 to 6 weeks postoperatively, the patient's cast is removed and an active exercise program with no resistance is initiated. A splint or bivalved cast is to be worn at all times except during periods of exercise. After 6 to 8 weeks the client progressively incorporates the use of the motor transfer into resistive exercise and function. At 8 to 10 weeks the day splint can be removed and continued to be worn for night only. Finally, after 6 months, the night splint can be discontinued.

CONCLUSION

The physician's, occupational therapist's, patient's, and family's expectations must be realistically and functionally related to the patient's level of motor control, coordination, and cognition. The physician and occupational therapist must work closely together pre- and postoperatively to discuss the various options and procedures available to improve function. The age when surgery should be performed must be determined on an individual basis, keeping in mind the procedure to be performed, the motivation of the patient, and the ability of the patient to carry out the treatment plan. Candidates for surgery should be at least 5 years old, however. By that age, the CNS has matured and children are old enough to participate in their own rehabilitation.[17,22–24] The patient and family members must be given a realistic explanation of what gains can be expected. With a carefully planned goal, the results of surgical procedures in the upper extremity can be gratifying for the patient.

REFERENCES

1. Hoffer MM. The use of the pathokinesiology laboratory to select muscles for tendon transfer in the cerebral palsy hand. *Clin Orthop Relat Res.* 1993;288:135–138.
2. Hoffer MM, Perry J, Melkonian GJ. Dynamic electromyography and decision-making for surgery in the upper extremity of patients with cerebral palsy. *J Hand Surg Am.* 1979;4:424–431.
3. Mowery CA, Gelberman RH, Rhoades CE. Upper extremity tendon transfer in cerebral palsy: electromyographic and functional analysis. *J Pediatr Orthop.* 1985;5:69–72.
4. Samilson RL, Morris JM. Surgical improvement of the cerebral palsied upper limb: electromyographic studies and results of 128 operations. *J Bone Joint Surg Am.* 1964;46:1202–1216.
5. Pinzur MS. Flexor origin release and functional prehension in adult spastic hand deformity. *J Hand Surg Br.* 1991;16:133–136.
6. Walters RL. Upper extremity surgery in stroke patients. *Clin Orthop Relat Res.* 1978;131:30–37.
7. Bleck E. *Orthopedic Management in Cerebral Palsy.* Philadelphia: J.B. Lippincott Co.; 1987.
8. Hunter JM, Schneider LH, Mackin EJ, Callahan AD. *Rehabilitation of the Hand: Surgery and Therapy.* 3rd ed. St. Louis, MO: C.V. Mosby; 1990.
9. Meals RA. *Hand Surgery Review.* 3rd ed. St Louis, MO: Quality Medical Publishing; 1990.
10. Zancolli EA. *Structural and dynamic bases of hand surgery.* 2nd ed. Philadelphia: J.B. Lippincott; 1979.
11. Goldner JL. Surgical treatment for cerebral palsy. In: Evarts C, ed. *Surgery of the Musculoskeletal System.* Edinburgh, Scotland: Livingstone; 1983:439–469.
12. Koman LA, Gelberman RH, Toby EB, Poehling GG. Cerebral palsy management of the upper extremity. *Clin Orthop Relat Res.* 1990;253:62–74.
13. Mital MA. Lengthening of the elbow flexors in cerebral palsy. *J Bone Joint Surg Am.* 1979;61:515–522.
14. Green WT, Banks HH. Flexor carpi ulnaris transplant and its use in cerebral palsy. *J Bone Joint Surg Am.* 1962;44:1343–1352.
15. Sakellarides HT, Mital MS, Lenzi WD. Treatment of pronation contractures of the forearm in cerebral palsy by changing the insertion of the pronator radii teres. *J Bone Joint Surg Am.* 1981;63:645–652.
16. Gschwind C, Tonkin M. Surgery for cerebral palsy, part 1: classification and operative procedure for pronation deformity. *J Hand Surg Br.* 1992;17:391–395.
17. Zancolli EA, Zancolli ER. Surgical management of the hemiplegic spastic hand in cerebral palsy. *Surg Clin North Am.* 1981;61:395–406.
18. Tonkin M, Gschwind C. Surgery for cerebral palsy, part 2: flexion deformity of the wrist and fingers. *J Hand Surg Br.* 1992;17:396–400.
19. McCue FC, Honner R, Chapman WC. Transfer of the brachioradialis for hands deformed by cerebral palsy. *J Bone Joint Surg Am.* 1970;52:1171–1180.
20. Hoffer MM, Lehman M, Mitani M. Long-term follow-up on tendon transfer to the extensors of the wrist and fingers in patients with cerebral palsy. *J Hand Surg Am.* 1986;11:836–840.
21. Pinzur MS, Wehner J, Kett N. Trilla M. Brachioradialis to finger extensor tendon transfer to achieve hand opening in acquired spasticity. *J Hand Surg Am.* 1988;13:522–549.
22. House JH, Gwathmey FW, Fidler MO. A dynamic approach to the thumb-in-palm deformity in cerebral palsy. *J Bone Joint Surg Am.* 1981;63:216–225.
23. Keats S. Surgical treatment of the hand in cerebral palsy: correction of thumb-in-palm and other deformities. *J Bone Joint Surg Am.* 1965;47;274–284.
24. Samilson RL. Principles of assessment of the upper limb in cerebral palsy. *Clin Orthop Relat Res.* 1966;47:105–115.

Occupational Therapy Department Guidelines Regarding Casting of Upper Extremities (UEs)

A. Occupational Therapy Staff will evaluate for and apply casts to patients' UEs on referral from RIC physician.

B. Types of casts applied include rigid circular, long arm, wrist, drop-out, platform.

C. Casts will be applied to manage abnormal tone and reverse contracture.

D. Application of casts is not recommended in the following:
  • Bilateral casts to patients with hypertension
  • Skin surfaces that are not intact
  • Joints with heterotopic ossification

E. Special consideration should be taken when casting patients with the following:
  • Heterotopic ossification
  • Decreased circulation/edema
  • Decreased sensation
  • Decreased orientation/alertness
  • Decreased stability of proximal joints

F. An Occupational Therapist experienced with casting will qualify staff to apply and monitor casts.

G. Qualified Occupational Therapy staff will assume responsibility for monitoring casts and documentation of process results and complications.

PROCEDURE

I. A. Referrals may be made to Occupational Therapy by any referring physician.

   B. Referral should include area to be casted (elbow, wrist, digits), extremity to be casted, and type of cast where possible.

   C. Any precautions which may affect the casting process. (See D Procedure)

   D. Where general Occupational Therapy orders made by physician, qualified Occupational Therapist may recommend casting and request orders as specified under B and C.

II. Qualifications of Occupational Therapy Personnel assessing for and applying cast.

   A. OTRs may be qualified to assess for and apply casts.

   B. To be qualified, therapist must be able to review the following information to qualified staff or resource clinician:
      1. indications for casting
      2. contraindications
      3. indications for various types of casts

Source: Copyright © Rehabilitation Institute of Chicago.

4. safety precautions/emergency removal procedures
5. cast application procedure

C. To be qualified, therapist must demonstrate cast technique for qualified staff/resource clinician by applying and removing cast on another staff member.

D. Cast applications, monitoring, and removal to be directly supervised by qualified therapist until technique skill is assured.

E. Prior to cast application, a qualified staff/resource clinician must okay type of cast for QA monitor.

F. OTIs may be qualified but must continue to be directly supervised in all aspects of casting.

II. Application of Casts

A. Precast evaluation to include
   1. goniometric evaluation
   2. notations of skin condition
   3. goniometric measurement of point-of-stretch reflex
   4. goniometric measurement of wrist position when hand is fully flexed, and fully extended
   5. record sensory status over area to be casted
   6. cognition
   7. spontaneous functional use
   8. use on command (arm placement/hand function)

B. Cleanse area to be casted

C. Apply stockinette

D. Apply felt over bony prominences and distal and proximal borders

E. Apply cotton padding

F. Apply plaster or fiberglass

III. Monitoring

A. Write date cast applied on cast and date to be removed.

B. After cast application, always monitor the following:
   1. red areas
   2. pulse at points distal to cast
   3. temperature comparison both UEs
   4. pain
   5. edema
   6. discoloration of hand/nailbeds
   7. dusky veins

C. Qualified Occupational Therapist will monitor casts with the above criteria hourly for 2 hours following cast application.

D. Following Occupational Therapy monitoring, Occupational Therapy will notify nurse in charge of patient that nursing should continue to monitor and schedule for monitoring.

E. If patient is to leave RIC with cast on, Occupational Therapy will instruct patient and/or family in monitoring procedures, provide written checklist, and schedule of monitoring.

IV. Emergency Removal

A. During regular Occupational Therapy hours, Occupational Therapy will be contacted to remove cast in event of complications.

B. During hours when Occupational Therapist is not available:
1. cast cutters can be obtained on 3rd floor cast room, and cast is removed by personnel trained in cast removal.
2. patient may be taken to nearest emergency room for cast removal.
3. if patient is to leave RIC with cast in place, Occupational Therapy will provide patient/family with name and phone number of nearest emergency room in writing.

IV. Documentation

A. Occupational Therapy will document in medical record (directly in patient chart) time and date of cast application and observations on hourly monitoring.

B. Occupational Therapy will document (in Occupational Therapy note) status prior to casting, changes in that status on removal of cast, and goals.

C. Occupational Therapy will enter status of cast into RICIS (order OT to RN communications).

APPENDIX B

Rehabilitation Institute of Chicago Physical Therapy Guideline

SUBJECT: CARE OF PATIENTS
TITLE: CASTING PERSONNEL QUALIFICATION

NUMBER:
DATE: 6/15/98
PAGE: 1 of 1

To ensure patient safety, physical therapists must demonstrate competency/casting qualification through the following procedure before they may apply casts independently.

THE PHYSICAL THERAPIST WILL:

- Review the following information with qualified personnel and/or through a self-learning package:
  - Indications and contraindications for casting
  - Indications for various types of casts
  - Safety precautions and emergency removal procedures
  - Cast application procedure
- Apply serial ankle and knee cast on qualified personnel.
- Fabricate footboard on qualified personnel (if applicable).
- Apply and remove casts (serial ankle, knee, and inhibitory footplates) on patient with direct supervision of qualified personnel.
- Several casts of the same type may need to be performed with direct supervision, preferably the same person.
- Once skill level is established, the therapist may independently apply and remove casts, but must have completed the competency in order to do so.

QUALIFIED PERSONNEL (CASTING OBSERVER):

- Demonstrates expertise with the casting technique for a minimum of 1 year of intensive casting experience (two or more cast applications per month)
- Supervises application, monitoring, and removal of casts until technical skill is assured.
- Fills out competency check-out and certificate of competency.

SUBJECT: CARE OF PATIENTS

TITLE: CASTING

NUMBER:

DATE: 6/15/98

PAGE: 1 of 1

It is the policy of RIC physical therapists to use lower extremity casting to maximize positioning and functional outcomes. To ensure patient safety, physical therapists will observe the following casting guidelines:

THE PHYSICAL THERAPIST WILL:

Determine Need for Cast
- Goals of Casting
  - Improve function
  - Increase ease of care
  - Improve positioning
- Evaluates patient for indications for a cast
  - Hypertonicity
  - Range of motion limitations which interfere with function or positioning
- Possible contraindications for casting
  - Heterotrophic ossification
  - Decreased circulation/edema factors
  - Severe tone
  - Agitation
  - Decreased orientation/alertness (*Extreme care and monitoring required*)
  - Decreased stability of proximal joints
  - Open skin area including chickenpox during breaking-out stage (*Casting only done to prevent further skin breakdown due to poor positioning. Cast applied and removed within 24 hours.*)

Obtain Referral
- Made by any physician, and should include:
  - Area to be casted
  - Extremity
  - Type of cast
  - Precautions affecting casting

Perform Pre-Cast Evaluation
- For Lower Extremity
  - Goniometric evaluation
  - Notations of skin conditions
  - Goniometric measurement of point of stretch reflex when appropriate
  - Sensory status of area to be casted
  - Motor function—functional use of the extremity
  - Obtain x-ray results to rule out orthopaedic contraindications

Apply Cast
- Process to be followed:
  - Cleansing of area to be casted
  - Apply stockinette

- Apply felt or cotton cast padding over bony prominences and distal and proximal borders
- Apply cotton cast padding to entire area
- Apply plaster or fiberglass
- Monitor position of joint(s) during procedure
- Write date of cast on the cast itself

Monitor Cast

■ The casted extremity should be monitored for the following indications of problems for 2 hours following the cast application and daily during treatment:
- Red areas
- Weak pulse at points distal to cast
- Temperature variance between casted and uncasted extremities
- Pain
- Edema
- Discoloration of hand/feet nailbeds
- Dusky areas
- Adverse behavioral responses

■ Notifies nurse to continue monitoring cast

Plan for Discharge from RIC or Follow-up for Outpatients

■ Patients and/or family must be given written instructions on monitoring procedures, a schedule for monitoring, and a letter to take to the emergency room to allow removal of cast if problems arise.

Plan for Emergency Removal

- The physical therapist or a trained physical therapist will be contacted during the day to remove a cast in case of complications.
- When a trained PT is not available, the resident on call (ROC) should be called.
- Cast cutters may be obtained from the 10th floor cast room to be used by trained personnel.
- Patients may be taken to the nearest emergency room for cast removal.

Document Casting Process and Results

■ Documentation directly in the medical chart must include the following:
- Status prior to casting
- Goals of casting
- Time/date of cast application and planned removal
- Observations during monitoring

■ Documentation of results of casting is included in regular PT progress and discharge notes.

Charging

Inpatient

■ Documentation directly in the medical chart must include the following:
- Charge "treatment I" for time spent during cast application and removal.
- Charge "equipment/fabrication charge" for time spent in the bivalving procedure.
- The PT who assists during casting does not charge, but accounts for time spent under "cast help" on the charge slip.

- Charge for casting materials through the equipment closet: fiberglass vs. plaster and BKA (ankle) vs. AKA (knee).

Outpatient

- Charge "Orthotic Fitting and Training" for time spent during cast application and removal.
- Charge "Orthotic Fitting and Training" for the time spent in bivalving procedure.
- The PT who assists during casting charges "Supervised Therapeutic Activity."
- Charge for casting materials through the equipment closet: fiberglass vs. plaster and BKA (ankle) vs. AKA (knee).

APPENDIX C

Rehabilitation Institute of Chicago Occupational Therapy Department
Upper Extremity Competency Check-Out

Therapist: _____   Date: _____

I. ORAL REVIEW

☐ Why are casts applied?
☐ What are two contraindications for application of a cast?
☐ What three factors may necessitate special consideration in application and monitoring?
☐ Name three types of casts and why you would choose that type.
☐ What type of cast would you choose for the following problems?
　–90° elbow contracture, fluctuating tone
　–90° elbow contracture, moderate, stable tone
　–no UE range limitations, patterned UE motion at elbow, wrist, and hand
　–elbow flexion, supination contracture, spasticity
　–elbow, wrist flexion contractures, wrist relaxes with elbow extension
　–mild finger flexion spasticity, full weight bearing possible, poor manipulation skills
　–wrist and finger flexor spasticity, no active wrist or finger extension, moderate flexor tendon and shortening
☐ What is the documentation procedure?
☐ When should casts be monitored?
☐ Name at least four things to be checked in monitoring the cast.
☐ What six areas should be evaluated prior to casting?
☐ What are maintenance options following casting and when should they be used?
☐ Discuss two articles related to casting.

II. APPLICATION PROCEDURE CHECK-OUT

☐ Pre-cast evaluation documented, physician's order, nursing notified.
☐ Preparation of materials.
☐ Water temperature.
☐ Size and application of stockinette.
☐ Size and application of padding and felt.
☐ Application of plaster bandages, direction, tightness, smoothness, tucking.
☐ Application of fiberglass bandages.
☐ Flaring, finishing.
☐ Dating and monitoring.

Source: Copyright © Rehabilitation Institute of Chicago.

III. REMOVAL PROCEDURE CHECK-OUT

- ❑ Explanation of cast cutter to patient.
- ❑ Determine line of cut.
- ❑ Holding of cast cutter and cutting.
- ❑ Use of cast spreader.
- ❑ Cutting padding.
- ❑ Changing blade.
- ❑ Bivalve technique.

IV. TYPES OF CASTS APPROVED

- ❑ Rigid circular elbow
            or
    Elbow drop-out, humeral portion enclosed
            or
    Elbow drop-out, forearm portion enclosed
- ❑ Long arm (rigid circular elbow, forearm, wrist)
            or
    Rigid circular wrist

EVALUATOR'S COMMENTS:

EVALUATOR: _____   DATE: _____

Rehabilitation Institute of Chicago Physical Therapy Department
Lower Extremity Casting Competency Check-Out

Serial Ankle P.T.:				
ORAL REVIEW	**DATE**			
State the rationale for application.				
State two contraindications for cast application.				
List three factors that may necessitate special consideration in the application and monitoring.				
Explain the documentation procedure.				
Name at least four things to be checked in monitoring the cast.				
Discuss maintenance options following casting and when they should be used.				
APPLICATION PROCEDURE CHECK-OUT				
Perform pre-cast evaluation, receive physician orders.				
Preparation of materials.				
Water temperature.				
Size and application of stockinette.				
Size and application of padding and felt.				
Application of bandages, direction, tightness, smoothness, tucking.				
Flaring, finishing.				
Dating and monitoring.				
Nursing notified.				
REMOVAL PROCEDURE CHECK-OUT				
Explanation of cast cutter to patient.				
Determine line of cut.				
Holding cast cutter and cutting.				
Use of cast spreader.				
Bivalve technique.				

EVALUATOR'S COMMENTS:

Approved By: _____ Date Passed: _____

Source: Copyright © Rehabilitation Institute of Chicago.

Rehabilitation Institute of Chicago Physical Therapy Department
Lower Extremity Casting Competency Check-Out

Serial Knee	P.T.:			
ORAL REVIEW	**DATE**			
State the rationale for application.				
State two contraindications for cast application.				
List three factors that may necessitate special consideration in the application and monitoring.				
Explain the documentation procedure.				
Name at least four things to be checked in monitoring the cast.				
Discuss maintenance options following casting and when they should be used.				
APPLICATION PROCEDURE CHECK-OUT				
Perform pre-cast evaluation, receive physician orders.				
Preparation of materials.				
Water temperature.				
Size and application of stockinette.				
Size and application of padding and felt.				
Application of bandages, direction, tightness, smoothness, tucking.				
Flaring, finishing.				
Dating and monitoring.				
Nursing notified.				
REMOVAL PROCEDURE CHECK-OUT				
Explanation of cast cutter to patient.				
Determine line of cut.				
Holding cast cutter and cutting.				
Use of cast spreader.				
Bivalve technique.				

EVALUATOR'S COMMENTS:

Approved By: _____   Date Passed: _____

Rehabilitation Institute of Chicago
Interdisciplinary Policy

SUBJECT: **CARE OF PATIENTS**

TITLE: **CAST APPLICATION AND REMOVAL**

NUMBER: **3.13.98**
DATE: **10/1/98**
PAGE: **1 of 1**

It is the policy of the Rehabilitation Institute of Chicago that therapists may use lower extremity or upper extremity casting to maximize positioning and functional outcomes. To ensure patient safety, all cast application and removal will be done by, or under the direct supervision of, a clinician who has demonstrated competence in casting. See discipline specific guidelines for competency testing, as well as standard guidelines for casting, patient selection, and evaluation.

Following cast application, casts applied on inpatients are monitored every shift for the first 24 hours and thereafter daily during therapy. Casts applied on outpatients are monitored following application until the cast is firmly set. The patient and/or caregiver are provided with instructions for continued monitoring at home and emergency removal procedures to follow. Casts on outpatients are then checked by the therapist during scheduled outpatient therapy visits.

Clinicians check for the following indicators of potential problems:

- Red areas
- Weak pulse at points distal to casts
- Temperature variance between casted and uncasted limbs
- Pain
- Edema
- Discoloration of hand/feet nail beds
- Dusky areas
- Adverse behavioral problems

In emergency situations, when an RIC therapist is not available, casts may be removed by the attending or resident MD or resident on call, or patients may be taken to the nearest emergency room.

At the request of/in consultation with the physician responsible for the cast application, the RIC Attending Physician may write orders for RIC staff to remove a cast applied by non-RIC personnel (eg, physician, cast room technician, OT, or PT). The RIC Attending Physician ordering cast removal or an attending physician designee must be available to assess the extremity following cast removal to assure appropriate management of any complications identified once the cast is removed.

Senior Vice President and Chief Operating
Officer, Hospital and Clinics

Senior Vice President, Medical Affairs and
Medical Director of the Institute

Vice President, Patient Care Services and
Chief Nurse Executive

APPENDIX E

Physical/Occupational Therapy Casting Program Worksheet

Name: _____   DOB: _____

Physician: _____   Diagnosis: _____

Circle   R   or   L   cast   Type: _____

Rationale for casting/goal:

_____

_____

_____

_____

Type of Cast Applied	Date On	Date Off	Problem Yes	Problem No	Comments
1.					
2.					
3.					

Date Met: _____   Discontinued: _____   Revised: _____

COMMENTS: (e.g., surgery, etc.):

_____

_____

_____

_____

Therapist Name   Date

APPENDIX F

Occupational Therapy/Physical Therapy Lower Extremity/Upper Extremity Casting Quality Monitor Log

NAME	IP	OP	DX	TYPE OF CAST	DATE ON	DATE OFF	SKIN PROBLEM		IF YES, EXPLAIN ANY ACTION TAKEN
							YES	NO	

Key: IP = Inpatient
        OP = Outpatient
        DX = Diagnosis

Source: Copyright © Rehabilitation Institute of Chicago.

APPENDIX G

Consent for Casting

I, _____, consent to being casted as prescribed by the RIC physician. The rationale and process have been explained to me along with my responsibility with monitoring and following my specific home program. Possible complications have also been reviewed with me.

Patient Signature: _____

Caregiver Signature: _____

Date: _____

Therapist: _____

APPENDIX H

Cast Care and Precautions

This cast will be removed on _____. Please read the following to learn about the care and precautions of your cast.

PRECAUTIONS

If your cast causes any of the following conditions, contact your therapist.

1. Swollen or puffy fingers/toes.
2. Differences in temperature or color between the casted and uncasted arm/leg.
3. Pain.
4. Numbness or tingling.
5. Blueness in fingernails/toenails.
6. Bad odor.

Do not get your cast wet since it will become soft and won't provide the same contact to the arm/leg. Cover the casted arm/foot with a plastic bag when taking a shower. If the cast becomes wet, it must be removed **as soon as possible** to prevent skin breakdown. (See handout related to emergency cast removal.)

Elevate your casted arm/hand periodically. Do not let it just hang at your side.

Check your cast_____.
                (therapist fill in frequency)

Therapist: _____

Telephone: (312) 908-_____.

Clinic Hours: 8:00–4:30 Monday through Friday

For weekend or holiday problems, go to the emergency room at
_____ Hospital, Phone _____.

APPENDIX I

Emergency Room Physician Letter

Date: _____

To Whom It May Concern:

I have applied plaster/fiberglass cast(s) to the _____, of _____, who is an outpatient at the Rehabilitation Institute of Chicago. The cast is being applied to gain range of motion. There is no fracture or joint instability to be concerned about if the cast is removed. The patient has been instructed to come to the emergency room if problems arise when we are not available.

Please remove the cast(s) if there is any question of compromised circulation or if the patient is complaining of significant pain. There is a minimum of three (3) layers of cotton padding under the shell of the cast.

If there are problems with the cast during the hours of 8:00 a.m.–4:30 p.m., please call me at _____.
                    Telephone number

Thank you for your assistance.

_____   _____
Therapist Name (please print)   Signature

                                                                 _____
                                                                 Date

Supplies and Equipment

LIST OF VENDORS BY TYPES OF MATERIALS AND SUPPLIES

The following list is a general guide to basic materials and supplies used for casting. Various choices of plaster and synthetic materials are available as well as casting accessories. As you become familiar with the properties and uses of the materials, you will probably develop preferences. The various product lines offer you many choices in the texture, modality, feel, and setting characteristics.

Splinting and Splinting Accessories

(Examples are Velcro, D-ring, Aquaplast, etc.)

North Coast Medical, Inc.
187 Stauffer Boulevard
San Jose, California
95125-1042
1-800-821-9319

Sammon Preston
P.O. Box 5071
Bolingbrook, Illinois
60440-5071
1-800-323-5547

Smith and Nephew Rolyan, Inc.
One Quality Drive
P.O. Box 1005
Germantown, Wisconsin
53022-8205
1-800-558-8633

Casting Supplies

(Examples are padding, cotton and synthetic stockinette, plaster, fiberglass, cast spreader, bandage scissors, cast scissors, casting tape, etc.)

North Coast Medical, Inc.
(see earlier listing)

Sammon Preston
(see earlier listing)

Smith and Nephew Rolyan, Inc.
(see earlier listing)

Johnson & Johnson Professional, Inc.
Orthopedic Division
325 Paramount Drive
Rayham, Massachusetts
02767
1-800-255-2500

Specific Products

J & J Specialist Fast-Setting Plaster
J & J Delta-Lite Conformable
Elasticon Elastic Tape
Zonas Porous Tape

3M Orthopedic Product Division
3M Center
Saint Paul, Minnesota
55144-1000
1-800-327-5380

Specific Products

3M Scotchcast Plus Fiberglass
3M Scotchcast Soft Cast

Carapace/DeRoyal
200 DeBusk Lane
Powell, Tennessee

Carapace Plaster
Cellona Plaster
Duracast Plaster
CaraGlas Ultra Synthetic Casting

Cast Cutters

Carapace/DeRoyal
(see earlier listing)

Johnson & Johnson Professional, Inc.
(see earlier listing)

Stryker
420 Alcott Street
Kalamazoo, Michigan
49001

(Distributor)
Alimed, Inc.
297 High Street
Dedham, Massachusetts
02026-9135
1-800-225-2610

You may want to contact your local hospital supply and orthopaedic supply vendors for information about the most recent casting products.

Index